CONTRIBUTIONS TO A PHILOSOPHY OF TECHNOLOGY

THEORY AND DECISION LIBRARY

AN INTERNATIONAL SERIES

IN THE PHILOSOPHY AND METHODOLOGY OF THE

SOCIAL AND BEHAVIORAL SCIENCES

VOLUME 5

CONTRIBUTIONS TO A PHILOSOPHY OF TECHNOLOGY

Studies in the Structure of Thinking in the Technological Sciences

Edited by

FRIEDRICH RAPP

Institut für Philosophie der Technischen Universität Berlin

D. REIDEL PUBLISHING COMPANY

DORDRECHT-HOLLAND / BOSTON-U.S.A.

Library of Congress Catalog Card Number 74–77969

ISBN 90 277 0433 3

Published by D. Reidel Publishing Company,
P.O. Box 17, Dordrecht, Holland

Sold and distributed in the U.S.A., Canada and Mexico
by D. Reidel Publishing Company, Inc.
306 Dartmouth Street, Boston,
Mass. 02116, U.S.A.

Printed in The Netherlands by D. Reidel, Dordrecht

TABLE OF CONTENTS

INTRODUCTION

The highly sophisticated techniques of modern engineering are normally conceived of in practical terms. Corresponding to the instrumental function of technology, they are designed to direct the forces of nature according to human purposes. Yet, as soon as the realm of mere skills is exceeded, the intended useful results can only be achieved through planned and preconceived action processes involving the deliberately considered application of well designed tools and devices. This is to say that in all complex cases *theoretical* reasoning becomes an indispensable means to accomplish the pragmatic technological aims. Hence the abstracting from the actual concrete function of technology opens the way to concentrate attention on the general conceptual framework involved. If this approach is adopted the relevant knowledge and the procedures applied clearly exhibit a logic of their own. This point of view leads to a methodological and even an epistemological analysis of the theoretical structure and the specific methods of procedure characteristic of modern technology. Investigations of this kind, that can be described as belonging to an analytical *philosophy of technology*, form the topic of this anthology.

The type of research in question here is closely akin to that of the philosophy of science. But it is an astonishing fact that the commonly accepted and carefully investigated philosophy of science has not yet found its counterpart in an established philosophy of technology. Hitherto only sporadic and widely scattered attempts have been made to investigate the structure of thinking in technology. It is the aim of this work to focus interest on this field by bringing together for the first time in book-form a collection of articles written by Eastern and Western scholars and pertaining exclusively to this subject. Logic and methodology being basically neutral to any *Weltanschauung*, possibly arising ideological divergences will be confined to the social implications of technology. As the state of the art does not allow for a systematic and embracing account, the papers included necessarily concern only selected topics and exhibit diversified approaches. Yet, taken together they seem to be able to give a survey of the prospects for this kind of inquiry by raising characteristic problems and giving provisional and tentative answers. A more complete view of the available literature and of works concerning specific topics can be gained from the select annotated bibliography on pp. 210–223.

Surely this lacuna in the investigation of technology cannot be due to the lack of importance of the subject, for it is hardly possible to overestimate the role of technology in our time. There must be other reasons. Obviously one of them is due to the *complexity* of the subject-matter, because in an analysis of technology one cannot disregard its social genesis and its actual function as easily as in the case of science. Being a synthesis of systematic knowledge and purposive action, technology requires a methodological analysis, which in addition to the relevant pattern of deductive reasoning also takes account of the pragmatic aims involved and the concrete actions performed. Yet this complex structure of technology is not only an obstacle to easy investigation. It also offers, at the same time, a chance to examine the much-disputed relationship between the logical and pragmatic approaches by means of an interesting and important paradigm, which in addition is likely to throw new light on the pragmatic aspect of science.

Furthermore, the *practical* approach of technology seems to be responsible for the rarity of reflections upon its theoretical structure. Being used to regarding a problem as solved as soon a solution has been found that 'works', the practitioner cares little about further scientific analysis. Thus only a part of the papers reprinted here has been written by engineers interested in a methodological analysis of their field; the majority of the contributions come from philosophers of science. Because they have to concentrate on practical tasks, engineers and technologists are normally not mindful of purely theoretical questions. Their main interest is in increasing the efficiency of the methods applied. In order to achieve this goal theoretical research is often carried out, too. But even in the more theoretical fields of cybernetics and systems engineering the intention generally is *to serve certain practical aims* and not to provide a better theoretical understanding for its own sake.

In contrast to this practical approach the inquiries of an analytical philosophy of technology are generally of no immediate use. They consist of investigation *about* the body of theories and the patterns of action relevant in designing and manufacturing technological objects. The results of such an analysis are necessarily too abstract and too general to be of direct concern to the professional engineer. The situation here is similar to that in the philosophy of science, which is also unable to give the scientist concrete advice for his research work. This of course does not mean that in *specific* fields, such as engineering education and design methodology, the conceptual analysis of technology cannot also lead to immediately useful results. Furthermore, the theoretical analysis may in the long run also have *indirect* practical consequences by providing a new understanding of the theoretical framework and the methods of procedure and thus influencing the further development of the discipline.

Moreover, the traditional view, which regards technology either as a sort of art or craft or at best as an *application of science*, may also have prevented a penetrating and comprehensive logical and methodological analysis. Indeed, if the pattern of knowledge and reasoning and the modes of procedure applied in technology could really be reduced to those of other fields, there would be no *specific subject* for the investigations of an analytical philosophy of technology.

But it is impossible to regard modern technology with its impressive results and its highly sophisticated methods as a mere craft supplemented by ingenious inspirations. The *scientific* influence in technology is quite evident; it consists in the application of the methods and findings of scientific research. And it is precisely the introduction of the systematic and scientific approach which stimulates interest in a methodological analysis of technology. Yet, the formula of applied sicence is an undue oversimplification. By no means all knowledge made use of in technology can be obtained from science, and not all scientific results are of relevance to technology. And modern science, in turn, depends on a high standard of technology which provides the apparatus and the instruments to produce the experimental conditions required to carry out the corresponding observations. As a result, science and technology are closely interrelated in many respects. Hence it will be one of the central tasks for an analytical philosophy of technology to reveal in detail the mutual dependence of the two areas, their common traits and their differences and thus to render their specific characteristics more evident.

The papers compiled here were originally published in quite *different contexts*. They can all be regarded as contributions to a philosophy of technology, though some of them were not explicitly intended to serve this purpose. Quite deliberately, studies written by *engineers* have been included. As their analysis is based on first-hand knowledge of the specfic problems and procedures of technology, it can serve as a useful counterpart to theoretically stylized schemes. Furthermore, the collaboration of engineers is necessary, as a philosophy of technology cannot simply omit the modern techniques of engineering design and systems engineering, although only the basic features and not all the details are relevant.

In order to limit the number of papers a certain *selection* had to be made. In doing this, the chief consideration was to refer to a broad sequence of relevant issues. Three of the items – those by Agassi/Wisdom, Skolimowski/ Jarvie and Pippard *et al.* – consist of *discussions*, which are reprinted in full, as it is the contrast of the positions, which is the most revealing. The latter comprises letters to the editor of the *The Engineer*. The other two as well as Bunge's article, are part of the stimulating symposium entitled 'Toward a Philosophy of Technology', organized by the *Society for the History of Technology*.

L. Tondl gives a general introduction to the philosophy of technology. After listing the main relevant problems he elaborates a definition of the concept of technology, outlines the main trends of technological progress, and analizes the relationship between the natural and the technological sciences.

The article by M. Bunge consists of an investigation of the theoretical elements of technology due to the application of scientific method. Bunge denies the validating force of practice as opposed to controlled research; his further discussion includes the relation of technological rule to scientific law.

The discussion between J. Agassi and J. O. Wisdom concentrates on the role of corroboration in science and technology, and on the chance for reducing invention to an algorithm. The authors differ especially in the assessment of the value of corroboration to science and technology (as derived from Popper's theory of falsifiability).

The discussion between H. Skolimowski and I. C. Jarvie centers upon the relationship of technology to science. Skolimowski discriminates both fields by examining their respective ideas of progress and investigates the specific thought patterns prevailing in various branches of technology. Jarvie points out that the way of thinking in technology depends on a social and traditional context.

F. Rapp explains how, in spite of their close interconnection, science and technology can be distinguished by means of their respective aims and procedures but not by referring to the difference between natural and artificial phenomena.

In his analysis, M. I. Mantell arrives at three different patterns of problem solving within the sequence of the scientific method: basic research, applied research, and systems approach; he also mentions the component elements of each pattern.

E. Jobst investigates the historical and the actual function of technology and science with regard to research and development as a source of technological progress.

D. Teichmann discusses and evaluates five different approaches to the classification of the technological sciences.

T. Kotarbiński examines technology in a broader and more elementary sense as that part of the theory of efficient action (praxiology) in which certain instruments and devices are applied.

M. Asimow's article gives an introductory survey of the basic features, problems, and principles of engineering design.

R. J. McCrory investigates in detail the structure of the design method and the various steps to be performed in order to arrive at satisfying results.

A. D. Hall elaborates and illustrates a general model of systems engineering by applying the morphological approach, i.e. by analyzing and describing a problem in terms of its basic variables.

The letters to the Editor comprise the discussion of various participants concerning the need for experimental work in engineering, the different types of experiment, their respective functions, and the complementary roles of theory and experiment.

The final article, by F. V. Lazarev and M. K. Trifonova, puts technology into the more embracing context of epistimology by analyzing the information properties of various types of apparatus as means of cognition.

Some further reflections on the subject-matter and the prospects of a philosophy of technology may be appropriate. Being the result of a *process of social action*, technology allows for an analysis from different points of view. For any of the manifold, closely interwoven components that are relevant in bringing about and making use of technological artifacts can be chosen as subject-matter for research. Applying a very rough classification, one could distinguish between two different approaches: (1) Attention may focus on the *logical and methodological structure* of the action processes performed, the knowledge applied within them and the objects thus realized. Investigations of this kind belong to the field of engineering, systems analysis and management. Normally they are of a pragmatic nature, designed to arrive at more efficient methods in research and development processes, or in planning, design and implementation. (2) The inquiry may concentrate on the *people involved* in the action process; this leads to the *social* aspect of technology. Here the main problem consists of giving an adequate account of the sources of technological change and of describing and evaluating its impact on human society.

With regard to philosophical inquiries the same distinctions hold. The logical and methodological analysis of the structure of technology can be distinguished from the philosophical investigation of its social impact. Thus the *philosophy of technology*, which forms the topic of this book, is characterized by a specific objectivizing point of view due to the disregard of the human element. Its counterpart consists in a far-reaching *technological philosophy*, not treated here, which explicitly refers to the people concerned in producing and making use of technical objects. While the first approach can be assigned to the field of a general methodology and theory of knowledge, the latter is part of a broadly conceived social and cultural philosophy.

Needless to say, the two aspects are complementary. Only together they can give an adequate and comprehensive view of the complex phenomenon of technology. Thus, for example, the analytical findings of the methodological analysis will also enter into philosophical inquiries about the social control of technological innovations or into discussions on problems of ecology. And even more speculative issues like reflections on the intellectual presuppositions which enabled the development of technology or discussions of the much-disputed question of the value of modern technology for mankind presuppose a former clarification of the relevant methodological problems.

Furthermore, the articles contained in this book generally do not take into account the *historical* development of technology. Although there are detailed descriptions of the specific traits of technology at different historical periods, a comprehensive and systematic methodological analysis is still wanting. But it should be kept in mind that in principle the history of technology is also open to an inquiry into the structure of thinking and the procedures prevailing at the various stages. Such investigations would surely be helpful in delimiting the 'historical contingency' of technology as opposed to its alleged 'inherent essence'.

However, one may doubt whether it is worth while to apply the fresh heading 'philosophy of technology' to a collection of problems which in principle could also be attached to *already existing* scientific investigations. Indeed, with some effort it would be possible to find a place for most of the questions raised here. Thus, for instance, the analysis of the action processes involved could be regarded as belonging to a general methodology of social actions; the formation and structure of technological knowledge and its theoretical formulation might be treated as part of the established philosophy of science; and the investigation of the production process and the structure of technical objects could be considered under the heading of systems engineering and cybernetics. But in this connection it should be kept in mind that the borders of scientific fields are the outcome of an historically contingent process of development. What really matters is the degree of understanding and systematization they yield. Following these criteria, it seems reasonable to collect the problems of a methodology of technology at least tentatively under a common title, as the patterns of thinking and action in the important field of technology exhibit so many specific traits which really deserve an investigation of their own.

Another suggestion might be that it would be better to *split* the vast field of an analytical philosophy of technology into more precise and detailed areas instead of treating it as a whole. As a matter of fact there are many special problems involved, such as the acquisition and application of technological knowledge, the stages of the production process and the preconceived structure of the objects thus produced, which could be investigated separately. Furthermore, the various branches of technology exhibit differences in procedure and in the levels of their theoretical elaboration. Yet, this situation is much akin to that that of the philosophy of science, where the unified – though necessarily simplified – approach of treating the different fields of science as having the same patterns has turned out successful. Likewise, the common traits of the various issues and branches of technology would seem to merit a unified analysis of their conceptual structure.

F. RAPP

ACKNOWLEDGEMENTS

The Editor and Publishers are grateful to the following for permission to reproduce copyright material: The University of Chicago Press for 'Technology as Applied Science', 'The Confusion Between Science and Technology in the Standard Philosophies of Science', 'The Need for Corroboration: Comments on J. Agassi's Paper', 'Planning for Success: A Reply to J. O. Wisdom', 'Rules for Making Discoveries: Reply to J. Agassi', 'The Structure of Thinking in Technology', 'The Social Character of Technological Problems: Comments on H. Skolimowski's Paper'; the American Society of Civil Engineers for 'Scientific Method – A Triad'; PWN-Polish Scientific Publishers, Warsaw for 'Instrumentalization of Actions'; Prentice-Hall, Inc. for 'A Philosophy of Engineering Design'; the American Society of Mechanical Engineers for 'The Design Method – A Scientific Approach to Valid Design'; the Institute of Electrical and Electronics Engineers for 'Three-Dimensional Morphology of Systems Engineering'; A. B. Pippard and *The Engineer* for 'The Role of Experiments in Applied Science'; *The Engineer* for two 'Letters to the Editor' by W. A. Tuplin and E. McEwen, and one 'Letter to the Editor' signed 'Your Reviewer'.

The Editor would also like to thank Mr. Ian J. Trotter for translating some of the articles into English. Permission to translate has been granted as follows: 'On the Concepts of 'Technology' and 'Technological Sciences' ' (L. Tondl and PWN-Polish Scientific Publishers, Warsaw); 'Technology and Natural Science – A Methodological Investigation' (the Editor); 'Specific Features of Technology in Its Interrelation with Natural Science' (VEB Deutscher Verlag der Wissenschaften, Berlin); 'On the Classification of the Technological Sciences' (VEB Deutscher Verlag für Grundstoffindustrie, Leipzig); 'The Role of Apparatus in Cognition and Its Classification' (Izdatel'stvo 'Vysšaja škola', Moscow).

L. TONDL

ON THE CONCEPTS OF 'TECHNOLOGY' AND 'TECHNOLOGICAL SCIENCES'*

The fact that the explanation and solution of philosophical problems is closely connected with science, is nowadays no longer a source of doubt for anybody. As a counterweight to the original, purely humanistic orientation of philosophy, in recent times the method connected with the results of the most important natural sciences has become current. Philosophers devoted their attention mainly to theoretical physics (more precisely, to some parts of it: quantum theory and the theory of relativity) and further, to astronomy, general biology and physiology and, to a lesser degree, to some other natural sciences.

In the analysis of certain basic philosophical categories, such as matter, movement, time and space, certain branches of the natural sciences attain greater or lesser importance. However, it has to be noted that, for example, the achievements of chemistry, biochemistry, biophysics and certain other sciences have remained as it were outside the sphere of interest of philosophy, not because they were not interesting or even not relevant from a philosophical point of view, but, above all, because the majority of philosophers underestimated the importance of such concepts as structure, organization, function, etc. And if hitherto philosophers have paid virtually no attention to chemistry, biochemistry, biophysics and certain other sciences, just as little, perhaps even less attention has been paid to the large field of technological sciences.

So for a long time it has been considered that those branches of science described as 'applied' cannot be connected directly with philosophical questions but only by means of those fundamental scientific disciplines which have been regarded as their theoretical basis. We think that this view is incorrect. Apart from the fact that the traditional differentiation between 'theoretical' and 'applied' sciences once promulgated by positivism is debatable (not only in the last few decades, but practically always, a number of theoretically important discoveries have been made in the so-called applied sciences), it is necessary to recognize the existence of directly philosophical problems of technology and the technological

F. Rapp (ed.), Contributions to a Philosophy of Technology, 1–18.

sciences and not only problems posed via physics, mechanics, etc. The results and successes of the so-called cybernetical technology of recent times have been especially convincing examples of this kind of problem. The existence of computing machines, self-regulating devices, complex automatic mechanisms and adaptive systems has led to questions which are of immense importance to philosophy. It would not be superfluous if the attention hitherto devoted only to certain narrow fields of technology and the technological sciences were also to be extended to the broad field of technology as a whole. It is certainly true that analysis of the process of knowledge and examination of such concepts as consciousness etc. have been based on the results of neurophysiology, psychology (and, unfortunately to a much lower degree, social psychology, linguistics etc.). But as a rule the role of technology, technological knowledge and technological sciences has not been taken into account, although this is not only important in the analysis of the concepts of practice but also in the determination of the relation between means and ends, the fixing of ends etc.

It is impossible to give a short outline of all the problems of the philosophy of technology and of the technological sciences. This can only be done on the basis of collaboration between specialists in technology and philosophy. It seems to us that the most important of these problems are:

(1) The analysis of the concept of technology, and this especially with reference to production, and to the forces and relations of production, in connection with the human activity of work, the classification of the basic means of technology etc.

(2) The explanation of the character of the technological sciences. In this connection we once again emphasize that the conception of technological sciences as merely 'applied' sciences is hardly tenable. In the field of the technological sciences basic research can exist alongside a broad spectrum of applied research. It would also be false to deny the cognitive function of the technological sciences. Cognition does not consist only in finding out *that* something exists within defined relations governed by natural laws, but also in finding out *how*, by making use of these relations, particular tasks can be performed.

(3) The relation of the technological sciences to the natural sciences and to the socio-economic sciences.

Hitherto great attention has been paid to the relation between the technological and the natural sciences and much less to the socio-economic sciences, especially the relations between the technological sciences and economics, although it is precisely this connection which is of exceptional importance.

(4) The explanation of certain basic concepts, in particular the concepts 'task' and 'to solve a task', the concepts of technological efficacy, technological norm, the concepts of technological and techno-economic criteria, etc.

In the following part of this paper we attempt to point out certain problems concerning the precise definition of the concept of technology which relate to the field of the first of the above-mentioned groups of problems involved in a philosophy of technology.

I. ON THE CONCEPT OF TECHNOLOGY

To the question of what technology is or can be, we can answer that technology consists of various instruments, machines, means of transport and communication, and measuring and experimental equipment; nowadays we also speak of agricultural technology, medical technology, computer technology, etc. It is evident that this enumerative approach is only an auxiliary means for explaining the concept of technology and can be neither complete nor exhaustive. Therefore besides such an extensional approach, the approach through the content is also justified. But here we already meet with the first difficulties.

Very often we meet with an explanation of the concept of technology which combines technology with material production. In simple terms, technology consists of the means of material production. But quite simple deliberation will convince us that such an explanation necessarily leads to doubtful consequences: in this case even the most primitive hammer of the Stone Age belongs to technology, whilst the telephone, television and the rocket certainly do not belong to technology as long as they are not used as means of material production.

In the *History of Technology* recently published in the U.S.S.R., Prof. Zvorykin gives such an explanation of the concept of technology in which he too emphasizes the connection between technology and material production. In his opinion technology consists of the means of labour which develop within the system of social production. Zvorykin stresses especially that the means of labour become technology only within the process of social production.

This explanation can, in my opinion, lead to a double interpretation: if technology becomes a creation of human labour only after being used directly in the social material production process, then medical, telecommunications, measuring and experimental technology are simply excluded from the concept of technology. Thus it is evident that such an interpretation, based on the identification of technology with production technology, is very narrow. In the other, broader interpretation, the connection between technology and the process of social production can be understood genetically and not functionally. In this case technology is what originates in the material production process and need not always and under all circumstances be used as means of material production.

It would seem that this broader interpretation is more acceptable, though there are also imperfections in it. First of all, this conception is limitless, because in it the concept of technology merges completely with the concept of material culture. For housing, clothing, food products and many other *things* are the results of material production. Therefore one cannot completely reject the functional point of view. But evidently it is not possible to reduce this functional approach exclusively to the tasks of production. There are also other tasks of human activity which in the process of the development of the human species became sundered from the original production tasks which are their basis and starting-point. But how can one nevertheless regard this functional point of view in a broader sense, so that one can separate the concept of technology from the broader concept of material culture?

In a certain sense the etymological term can help us. The word 'technology' is derived from the Greek τέχνη, which means skill, art. This is, strictly speaking, what one could call the 'subjective, purposive moment' of technology. In order to explain this moment more precisely, we turn to the classical characterization of the essence of man which Karl Marx gave and which develops a remark by Franklin about man as a 'tool-making animal'. Man produces instruments (and also machines of all kinds and technical equipment which becomes more and more complicated and expensive) in order to reach certain preconceived goals, in order to realize his intentions, plans, etc. Thus, man first creates the final result of his activity ideally, as the image of his creation, as a goal, a plan. But what is important and essential here is not only that this is not a passive image of the planned or intended result, but also that it relates to consequentiality, to the appropriateness of the means to the given goal and the manner of their use; in short: the ideal image of the results includes not only the '*what*' but also the '*how*'. And precisely this '*how*' is especially important for the elucidation of the subjective moment of technology. Of what use are the most efficient and perfect means if we do not know for what goals and how we can use them. Thus we cannot limit the concept of technology to the concept of the means of human activity taken as a whole without taking into account the level of technological knowledge, the degree of control of technological processes etc. One can also add that in every historical society there has always existed a certain hierarchy of what we have expressed by the words '*what*' and '*how*'. There always exists a certain scale of goals, the basis of which consists of those goals which are connected with the sustenance of human existence. Therefore the needs of social production also constitute the basis of the development of technology. But technology on every level of its development itself creates new production needs. The need to accelerate the transport of people and goods was the basis for the creation and improvement of railways, cars, aeroplanes, etc. But at the same time the existence of these and other means of transport technology gave rise to such social needs as were connected with it (maintenance, further improvement, etc.).

The subjective moment of technology is thus closely connected with two factors:

(a) the purposive character of human behaviour, in which the decisive criterion is the final result of this behaviour,

(b) a certain knowledge of appropriate procedures, of appropriate technological consequentiality (in the widest meaning of this word).[1]

So in the explanation of the concept of technology we cannot ignore the subjective moment but at the same time it is by no means sufficient for our purposes.

We shall not succeed on this basis either in distinguishing more clearly the concepts of technology and material culture. The main emphasis must therefore be laid on the objective moments. To characterize these objective moments of technology one single definition is not sufficient. We shall therefore give some definitions of this kind which of course must not be considered in isolation but in their relations to one another.

(1) The most general characterization of technology can be given in the following way: technology is everything which man in his activity *puts*

between himself and the objective world and its individual parts with the goal of transforming this world in accordance with his needs and intentions. The concept 'to transform the objective world' must here be understood in a broad sense: to change any properties of the phenomena of the objective world, including, for example, spatial, temporal and other properties.

(2) Another characteristic trait of technology is that it comprises the sum of the *resources which increase the efficiency* of human activity. Here the concept of efficiency, too, must be considered in a more general sense, not only in its physical, energetic sense.

Man can increase his efficiency, i.e. achieve better results, while expending the same or even a lesser amount of his own force, in the most varied manner. Above all, it is possible to increase the energetic capacities of man: man cannot break a log into pieces with his bare hands, but the axe helps him to do so. He cannot raise heavy tree-trunks without the help of a lever, etc. (Thus one can enumerate all the so-called classical tools.) Man as a source of energy and as a driving force evidently has very limited capacities. But the strength of the human hands can be increased many times in the most varied ways: by the use of means which directly increase the efficiency of the physical activity of man, by employing in conjunction with the tools of labour sources of energy outside the human organism which man applied and still applies (the strength of animals, the energy of wind and water, heat energy and the power of heat engines in general, and of various transformations of energy, etc.). The Soviet logician G. N. Povarov calls this sphere of technology that of production or work.

Besides this sphere, there is also a field concerned with raising the efficiency of the human senses, and not with increasing the efficiency of human physical activity. Povarov speaks in this context of instruments of the senses and instruments of perception. The telescope, the microscope and a whole series of other instruments are also components of technology, but ones which increase the efficiency of human sight and hearing, not of the muscles.

Another sphere of technology comprises the means which increase the efficiency of the intellectual activity of man. These are not only the equipment which we know from the most recent developments in technology. For thousands of years man has raised the efficiency of his intellectual activity by making use of aids to memory and communication. The ability of the human brain to retain information and to recall it at a certain time and in a certain place has thus been broadened. Aids to memory and communication have basically one and the same function: the transmission and storage of information, in the first case in time and in the second in space. So the aids to memory and communication are not the product of modern times, although recent years in particular have seen far-reaching revolutionary changes in the development of these resources.

Printing, photography, telegraphy, broadcasting, television, computing machines, etc., all these are nowadays an important component of technology. But all these aids have a great and important history, which begins with the first primitive written notation and calculating machines and leads to modern computing machines. The efficiency of intellectual activity is also increased by devices capable of decision-making on the basis of data which they themselves procure. The ability of decision-making means here choosing from two or more possible alternative procedures. This sphere of technology includes the wide field of automata, beginning with the simplest devices, such as for example, the cross-bow. It is evident that at present we are only on the threshold of a far-reaching development of this field of technology concerned with broadening and deepening the efficiency of the control, decision-making and other intellectual functions of man.

(3) To characterize the concept of technology one can also point to the following important trait, which is based on the concept of causality. In this connection we shall refer to some ideas of the Soviet mathematician A. A. Markov: technology is always a certain *synthesis of causal nets*, implemented by man with the aim of attaining certain desired, planned results.

In nature a great number of changes take place which can be explained causally, which can be understood as causal nets. These changes take place independently of man, his will, and his plans. In this sense nature is constantly changing and reproducing itself. But man is able to create new causal nets and in so doing, strictly speaking, create a 'new nature'. Of course, he can create this new nature only by making use of the objectively existing causal relations, by effecting a synthesis of the new nature from that nature which really exists and which is already known to him.

But the idea of a synthesis of causal nets must be expressed more precisely in some respects.

First of all, it is necessary to point out that in the synthesis of causal nets which he afterwards uses to bring about changes in nature, man can base himself on various spheres of nature, on various forms of the motion of matter.

The basis and starting-point of technology is thus not only the mechanical and physical movement of matter. Scientific and technological progress will evidently lead in the future to the so-called higher forms of the motion of matter playing an ever-increasing part as the basis of technology. Hitherto the overwhelming majority of the tasks which technology fulfils, have been achieved by means of a synthesis of elements which have physical character or, more precisely, the physical properties of which are made use of (i.e. thermal, mechanical, electromagnetic properties, etc.). But this does not exclude the possibility of a synthesis of technical devices from elements in which other properties too are made use of, mainly chemical and bio-chemical ones, etc.

Furthermore, it is necessary to remember that synthesis presupposes analysis. Every synthesis of elements presupposes that to a certain degree we know their causal relations and their properties. In other words, every synthesis which one can call technology presupposes a certain level of knowledge. It is evident that the more complex the task of synthesis is, the higher the level of knowledge must be. Therefore modern technological progress is, strictly speaking, impossible without science, without extension of scientific knowledge. Science, and especially natural science, have a double meaning for the formulation and achievement of the tasks of synthesis: on the one hand they indicate which tasks are possible, and, on the other hand, what are the limits of these possibilities, which tasks can be solved and which not by making use of elements of a certain kind.

As already mentioned, it would be wrong to connect the idea of a synthesis of causal nets only with the simplest forms of the movement of matter, only with the physical properties of the elements used in the synthesis. A necessary condition for every synthesis is the interaction of two or more systems between which a certain exchange takes place. This exchange can involve all three fundamental levels of exchange: those of matter, energy and information. In the case of turning metal on a lathe the most important aspect is obviously a kind of exchange of matter. In such technological machines as heat engines, generators of electrical energy, etc., a certain energy exchange takes place. What takes place in storage, communication, and decision-making devices is primarily a specific kind of information exchange. At the same time it is necessary to note that the three levels mentioned cannot be isolated. They can only be isolated in abstraction. Therefore such concepts as 'matter' ,'energy' and 'information' do not express absolute properties but always only properties which manifest themselves in a certain interrelation, interaction and mutual influence of two or more systems.

From this point of view one can describe the general function of technology as an exchange of matter, energy or information *organized by man*. The concept of organization in its modern (i.e. cybernetic) sense is here evidently of fundamental importance: man strives to raise the level of organization of systems by means of technology. From this follows a concept which one might characterize as the *antientropic function of technology*. By means of technology man reduces the entropy of real systems. Of course this definition can have only a relative meaning, i.e., it refers only to relatively isolated systems.

(4) The concept of technology can be characterized by defining, at the various stages, the function of man in the tasks performed by technological equipment. Thus one can distinguish various types of technological devices. This point of view is contained – in a simple and to a certain degree naive form – in the question: what does man do and what do the technological devices do in his stead? Of course this question simplifies the point of view mentioned, if for no other reason than that it does not distinguish what man does with respect to the direct operation of a technological device and what he does in the way of preparation or programming of this operation. Nonetheless, this point of view (we have in mind the direct operation of a technological device) can be a rough means for differentiating the various types and, at the same time, the historical stages of technology.

The capacities of technological resources, translated into human activity, are increased – roughly speaking – as the proportion of what man does on his own in the direct operation of the technological device decreases and the proportion of what the machine does in his stead increases.

In this not too precise statement the term 'machine' is applied in the broad meaning of technological device in general. As a rule, the term 'machine' is used in three different senses: (a) in the sense of the so-called physical machine (mechanism) (for example: lever, wedge, inclined plane, etc., i.e. in the sense of a device which transforms mechanical energy; in this sense even the simplest hoe is a machine); (b) in the sense of the so-called classical machine, i.e. a device combined with some exterior source of energy outside man and performing the required operations; (c) in the sense of a technological device in general. Besides this, the term 'machine' can be applied in an extremely abstract sense. At this point we come to a certain concept, i.e. the conception of a technological device, in the most general meaning of the word, which to a certain degree effects a purposive exchange of matter, energy or information, planned by man (this device itself being in turn a product of human activity), which permits man's share in the indirect operation of the technological device to be classified as follows:

(a) man himself constitutes the source of energy, the motive force, and the source of information. A technological device of this kind man sets in motion by the force of his muscles. In this case we usually speak of *tools*. Such tools are, for example, a knife, an axe, a screw-driver, a chisel, etc. Man operates a tool by putting it between himself and individual objects in nature with the aim of transforming these objects, adapting them to his needs, attaining certain desired results etc. Most tools operate in such a way that man must guide and control them directly, i.e., he himself is the source of information.

(b) Man can also operate these tools or certain organized systems of such tools by making use of external sources of energy outside himself: the tractive force of animals, the force of wind, water, etc. The higher devices of this type are based on such kinds of energy conversion as do not take place within the limits of a single form of the motion of matter but are based on the transformation of one form into another. As is well-known, in this respect, the steam engines brought about revolutionary change. Nowadays steam engines constitute a particular field of technology, based on the transformation of various forms of the motion of matter. An analysis of this kind of technology, the so-called *classical machine*, was given by Karl Marx in his *Capital*. Marx showed that in these classical machines there is always a source of energy, a system of transmission, and, finally, a system of tools which perform certain operations. It is necessary to add that these machines always require the presence of man to guide and control their functions. Thus for these machines man is no longer the source of energy, but he is still the source of information.

(c) If we replace man's share in the work of the machine by a technological device, we pass from the classical machine to the *automaton*. The automaton differs from the classical machine of the classical type by being capable of decision-making and of controlling itself. In other words, the automaton operates on the basis of its own information. Of course, such a sudden jump from the machine of the classical type to the automaton exists only in the abstract. With actual technological devices one can only speak of a greater or lesser degree of automation. For example, Watt's centrifugal governor in the steam engine is an elementary automatic device of this kind, capable of controlling itself. On the other hand, even the most perfect computing machines are not complete automata, because they always operate on the basis of a certain programme prepared by man (and depending on the level of automation attained). In this connection attention is due to the endeavour to achieve higher types of automation, especially in the form of 'learning machines', 'self-organizing systems', etc., which represent higher forms of programming.

The three concepts: 'tool' – 'machine (of the classical type)' – 'automaton' – cover, in a simplified and abbreviated form, the basic types of technology and are at the same time distinct historical types. But once again it must be stressed that, really, these types exist in an absolutely pure form only in the abstract. In practice it is of importance to distinguish that sphere of technology which replaces the physical efforts of man (the sphere of mechanization), and that sphere which replaces the intellectual functions of man, the functions of control and decision-making (the sphere of automation).

If it is possible, as we have already noted, to combine the concept of technology with intentional human activity aimed at increasing the degree of organization in the exchange of matter, energy and information, then it is also possible to indicate the principal trends in technological progress:

(1) In the first place there are problems involving *materials*. The principal prerequisites for increasing the degree of organization in the exchange of matter are the discovery of new properties of materials already known and acquiring more precise knowledge of these properties. (To this group of problems belong, for example, the problems of elasticity and strength, investigation of the properties of steel, concrete, etc.) We are, however, also concerned here with the creation of completely new materials, such as, for example, plastics, the investigation of polymers and the synthesis of

materials with properties much more suitable for new tasks of technology than natural substances.

(2) In the second place there are *energy* problems, the problems concerning technically useful transformations of different forms of energy. These include the broad complex of problems of raising the efficiency of energy conversions already known, the problems of new, more efficient sources of energy (especially the direct energy exchange, which does not make use of mechanical elements with rotating parts, and also conversions of chemical and thermal energy into electrical energy, the problems of controlled thermonuclear fusion, etc.).

(3) Finally there are problems of the *exchange of information*, the problems of control and controlled systems, and of the automation of various tasks which presuppose the use of algorithms and the programming of these tasks for the technological devices concerned, and the vast problem of technological modelling as a means to economize and optimize the solution of various tasks.

This differentiation of the main trends of technological progress, which are, of course, closely related, indicates at the same time the basic perspectives of the further development of modern technology and its various specialized fields.

II. THE NATURAL AND THE TECHNOLOGICAL SCIENCES

Very often the technological sciences are regarded as an 'application' of certain basic natural sciences, especially physics. This conception was introduced by the older positivism, and it is contained in a number of positivist classifications of sciences. It is not quite clear what is to be understood by 'application' in this view. From the historical point of view the problems of technology are, as a rule, older than those of the natural sciences. It was precisely the solution of tasks of a technological character which gave a strong stimulus to the growth of knowledge in the natural sciences.

The distinction between natural and technological sciences is a product of the increasing division of labour in science. But the fact should not be overlooked that the connection with technology was and still is a particularly important source of impulses for the natural sciences. Hence the traditional division into 'theoretical, natural sciences' and 'applied sciences' cannot be regarded as satisfactory from a present-day point of view. We do not want to underestimate the traditional problem of the classification of sciences, but it must necessarily be regarded as a problem derived from others in two respects: (a) derived with respect to a differentiation of the various 'forms of the motion of matter', a differentiation of the various 'levels' on which one can investigate the objective world; (b) derived with respect to the division of labour in science and the demands of optimal organization of this division.

If we retain only point (a), be it in Engels' original version or in the more precise formulation of it as given, for example, in the works of B. M. Kedrov, or, finally, on the basis of the 'theory of levels', established in the works of D. Bohm, J. P. Vigier and others, we would unavoidably be faced with a whole lot of difficulties, both in defining the place of the technological sciences and in defining the term 'application' more precisely. If we turn to point (b), then we shall see that the present division of labour in science is based on the difference between 'basic research', concerned with long-range goals going far beyond the present possibilities of practical use, and 'applied research', which is oriented towards the possibilities of practical exploitation. (This distinction involves a number of questions which are not the concern of this paper.) From the point of view of the division of labour in science organized as described here, it might seem absolutely natural that the technological sciences should be concerned with 'applied research'. But in fact the traditional division into 'theoretical' and 'applied' sciences and the differentiation between 'basic' and 'applied' research are not identical and overlap only partially. As far as the technological sciences are concerned, the following points cannot be ignored:

(1) The technological sciences today comprise a very broad and manifold range of special fields, the differences between some of which are greater than those between related natural and technological sciences. In many cases (for example with the semiconductors), it is very difficult to decide whether a given research project belongs purely to the natural sciences or to an indivisible unified complex of science and technology.

(2) It is not possible to assign all that is covered by the concept of the technological sciences to the field of applied research. Some parts of the technological sciences can have the character of basic research, whilst

others belong to applied research. The same thing, though not to the same degree, also applies to some natural sciences.

(3) If we regard the technological sciences merely as 'application', in the sense of making use of knowledge already available to solve particular tasks, then in this case a creative character could only be ascribed to the natural sciences. In reality, creative and non-creative elements are intermingled in both the natural and the technological sciences.

The term 'creative work' in science can be interpreted in different ways; it is possible to associate with it an evaluative or an emotional point of view. But when we speak of the interconnection and intermingling of creative and non-creative elements in scientific work, what we have in mind is something different, namely the methodological meaning of the term 'creative work'. In order to make this meaning more precise, we must first explain the concept 'to solve a task',

In both the natural and the technological sciences the concept 'to solve a task' can cover operations and procedures of very different character: measurement, predictions, scientific explanations, technological and construction projects, etc. 'To solve a task' means to divide it into a series of tasks which it is possible to regard as solved. From this explanation the connection between the concept 'to solve a task' and the concept 'algorithm' is evident. Hence one can say that scientific work has algorithmic as well as non-algorithmic character, that it includes creative as well as non-creative elements. The main purpose of science is to formulate new tasks, to look for new solutions to these tasks and to ones already known, and also to look for better solutions to previous tasks. Usually one endeavours to make use of the method found and the final solution as guidance for the solution of tasks of a similar or analogous type.

In science all three basic groups of tasks are found, tasks which are totally insoluble or cannot be solved in a purely algorithmic way (these concepts are not always identical, for there can be tasks which cannot be solved in an algorithmic way but to which isolated solutions can be found), tasks as yet unsolved and tasks already solved. The difference between the natural and technological sciences does not consist in the fact that in them only one of the three groups of tasks mentioned predominates, but in the fact that on the one hand we have in view the *possibility* of finding a solution (in the natural sciences) and on the other hand the *implementing* of the solution (in the technological sciences).

A related problem is that of *determining the limits and the limiting possibilities* and explaining these limits by means of science. Above all, it is necessary to note that stressing the limits of the possibilities of fulfilling a particular task has nothing in common with agnosticism or fundamental scepticism. Just the opposite: when we know the limits of the possibilities of technology we can economize on human effort in achieving progress in science and technology. The assertion that a heat engine with 100% efficiency is fundamentally impossible (in more simple terms: there cannot be a transformation of energy with 100% efficiency) is not agnosticism, on the contrary it has saved a great deal of useless effort.

The problem of limits has two meanings: in the natural sciences it always consists in the exact determination of the limits of a certain action, and of the ultimate properties of a certain process, and this always means progress in theoretical knowledge. Only if we know the exact limits of a certain relationship, do we know its precise meaning. Precise knowledge of the location of this limit means in technology knowledge of the limiting possibilities for technical constructions which any technological implementation can more or less approach, but never completely attain, let alone exceed.

This is also the meaning of the most basic laws of nature, the laws of conservation. As is well-known, only the exact explanation of the law of the conservation and transformation of energy proved that the efforts of many generations to create a *perpetuum mobile* were not feasible. In the works of Carnot, based on theoretical experiments with the *ideal steam engine* and on the laws of thermodynamics, the construction of heat engines with 100% efficiency was shown to be impossible.

In explaining the role of the limits and limiting possibilities in the interrelations of natural science and technology it is not necessary to refer only to the nineteenth century. Discoveries in natural science in the twentieth century are of similar importance. One of the presuppositions of the special theory of relativity, formulated by Einstein and based on the negative result of the well-known *Michelson-Morley experiment*, was the hypothesis of the limiting velocity which cannot be exceeded by moving particles.

The so-called relativistic effects were only proved much later, but despite this the original transformation formulae introduced by Einstein have this same meaning of a limiting velocity, which cannot be exceeded even in the most sophisticated apparatus of present-day physics.

Heisenberg's well-known *uncertainty principle* also expresses certain limits of possibility, for example concerning the possibility of representing a microprocess by macroscopic means (by experimental or measuring devices as well as by theoretical models corresponding to the macroscopic world). Today we know that *Heisenberg's principle* can be generalized in a certain way, for example, in information theory through the relations between a transmitter of information and a channel of a certain capacity which never is and never can be completely unlimited. Finally, *C. Shannon's fundamental theorem of information theory* refers basically to certain limits of the transmissibility of information, which it is impossible to exceed by any technological means.

Not only the results of natural science, but some important achievements of mathematics as well point to certain possibilities of technological implementation. There is, above all, the famous *theorem of Gödel* about the limits of formalization, based on the proof of the so-called formally undecidable theorems in formalized systems of a certain level. This theorem brought about a revolutionary change in the hitherto prevailing views on the foundations of mathematics. The impossibility of formalization (the limits of formalization have been proved in the work of Gödel, Church, Turing, Markov, Novikov and others) also has a technological meaning: what cannot be formalized cannot be implemented in a technological device.

In a certain respect the possibilities of technology are expressed more precisely by the concept '*Turing machine*,' which is more exact than the formulation of the original intuitive concept of algorithm. Here one can see a certain analogy. The *ideal steam engine of Carnot* never existed and could never exist, never became a technical reality; in the very same way, the Turing machine could never become a technical reality, as it presupposes the infinity of the so-called external memory store (the infinity of the 'tape' of the *Turing machine*). But despite all this, the *ideal steam engine of Carnot* contributed much more to precise analysis of the work of the heat engine than, as F. Engels remarked, the inductive investigation of thousands of steam engines. The ideal construction of the *Turing machine* has a similar importance with respect to devices for information-processing, or more precisely, with respect to the algorithmic characteristics of the way they function.

If we consider the examples adduced, which refer to the discovery of limits and limiting possibilities we can sum up certain characteristic traits of the natural and the technological sciences in the following way.

The main interest of the *natural sciences* is concentrated on the following tasks:

(1) They aim at revealing the various laws in the 'behaviour' of systems of very heterogeneous character; these systems can be:

(a) real, existing in nature;
(b) not as yet detected in nature, but assumed to be objectively possible;
(c) created by man;
(d) abstract models of the systems of types (a), (b), (c).

(2) The concept of 'behaviour' generally has the character of a dispositional predicate. In the language of the natural sciences this character can be expressed in very different ways, for example, by referring to the conditions which do or do not bring about a certain kind of 'behaviour', by referring to causal relations, in formulations of the type 'if ..., then ...', etc. For example, under conditions of normal pressure, water between the temperatures of 0 °C and 100 °C is in a liquid state, it 'behaves' as a liquid. The concept of 'behaviour' can also be explained in the terminology of the cybernetic approach: By 'behaviour' we understand the pattern of the output sequence as dependent on the pattern of the input sequence.

(3) An important task of cognition in the natural sciences is precise determination of the limit of the possible 'behaviour' of a system under given or postulated conditions. The important role of the 'nodal points' in nature, where the 'behaviour' of a system changes qualitatively, was already stressed by Engels. These limits of 'behaviour' or 'nodal points' also function as qualitative limits between different fields of nature and at the same time as qualitative transitions from one field to another.

The main interest of the *technological sciences* is concentrated on the following problems:

(1) Investigating the realization of the synthesis of a certain 'behaviour' from given elements. It is not necessary to draw special attention to the fact that the original model of the 'behaviour' to be synthesized consists of certain kinds of human behaviour relating above all to productive activity, or of the 'behaviour' of other living or non-living systems. So man originally imitated nature by technological means (and also

himself, as a part of nature); he imitated the 'behaviour' of natural systems, so that he synthesized this or a similar kind of 'behaviour' with the aid of the means he was capable of creating.

(2) The investigation of the ways and means of optimal synthesis of the particular 'behaviour' from given elements, the possibility of economizing on the synthesis and the possibility of optimizing the conditions for the 'behaviour'.

(3) But the technological sciences do not restrict themselves only to the 'imitation of nature' and the synthesis of 'behaviour' already known. They can create a 'new nature', they can synthesize a new kind of 'behaviour', not existing as such in nature, the results of which can fulfil social needs to a higher degree. Tasks of this kind are always circumscribed by the specific limits existing for them. In such cases of new synthesis, one can speak as a rule, of a maximal approximation to these limits.

In a certain sense some basic tasks of the natural and technological sciences are coming closer and closer to each other, intermingling and even becoming identical: if in our times the natural sciences develop in such a way as to switch over from the investigation of the 'behaviour' of a system under original conditions (ones that are 'normal', within man's reach, in nature as formerly or still accessible to man, under the existing conditions) to the investigation of 'behaviour' under 'extreme' conditions (for example, conditions of high pressure or high temperature, high energy, conditions of the 'fourth state', etc.), then it is one of the tasks of the technological sciences to induce these 'extreme' conditions. From this point of view progress in the natural sciences is impossible unless the most modern technological means find access to all fields of investigation, unless there is close interaction between the natural and the technological sciences.

Prague

NOTES

* Originally published in Czech under the title 'Slovo o filosofii techniky', in *Filosofický časopis* **12** (1964), 281–293 and in Russian under the title 'O ponjatijax 'texnika' i 'texničeskie nauki' ' in *Organon* **3** (1966), 111–125.

[1] [At this point a paragraph has been omitted which deals with a further meaning of the Russian term 'texnika' (mainly used here in the sense of 'technology'). Since English has a separate expression for this further meaning ('technique'), the problem dealt with at this juncture does not arise for the English reader.]

M. BUNGE

TECHNOLOGY AS APPLIED SCIENCE*

The application of the scientific method and of scientific theories of the attainment of practical goals poses interesting philosophical problems, such as the nature of technological knowledge, the alleged validating power of action, the relation of technological rule to scientific law, and the effects of technological forecast on human behavior. These problems have been neglected by most philosophers, probably because the peculiarities of modern technology, and particularly the differences between it and pure science, are realized infrequently and cannot be realized as long as technologies are mistaken for crafts and regarded as theory-free. The present paper deals with those problems and is therefore an essay in the nearly non-existent philosophy of technology.

I. SCIENCE: PURE AND APPLIED

The terms 'technology' and 'applied science' will be taken here as synonymous, although neither is adequate: in fact, 'technology' suggests the study of practical arts rather than a scientific discipline, and 'applied science' suggests the application of scientific ideas rather than that of the scientific method. Since 'techniqe' is ambiguous and 'epistechnique' unborn, we shall adopt the current lack of respect for etymology and go over to more serious matters.

The method and the theories of science can be applied either to increasing our knowledge of the external and the internal reality or to enhancing our welfare and power. If the goal is purely cognitive, pure science is obtained; if primarily practical, applied science. Thus, whereas cytology is a branch of pure science, cancer research is one of applied research. The chief divisions of contemporary applied science are the physical technologies (e.g., mechanical engineering), the biological technologies (e.g., pharmacology), the social technologies (e.g., operations research), and the thought technologies (e.g., computer science). In many cases technology succeeds a craft: it solves some of the latter's problems by approaching them scientifically. In other cases, particularly those of the social and thought technologies, there is no antecedent prescientific skill because the problems themselves are new. But in every case a distinction must be made between artisanal knowledge and scientific knowledge, as well as between pure research, applied research, and the applications of either to action.

The division of science into pure and applied is often challenged on the ground that all research is ultimately oriented toward satisfying needs of some sort or other. But the line must be drawn if we want to account for the differences in outlook and motivation between the investigator who searches for a new law of nature and the investigator who applies known laws to the design of a useful gadget: whereas the former wants to understand things better, the latter wishes to improve our mastery over them. At other times the difference is acknowledged, but it is claimed that applied science is the source of pure science rather than the other way around. Clearly, though, there must be some knowledge before it can be applied, unless it happens to be a skill or know-how rather than conceptual knowledge.

What is true is that action – industry, government, warfare, education, etc. – often *poses problems* that can be solved only by pure science. And if such problems are worked out in the free and lofty spirit of pure science, the solutions to them eventually may be applied to the attainment of practical goals. In short, practice is one of the sources of scientific problems, the other being sheer intellectual curiosity. But giving birth is not rearing. A whole cycle must be performed before anything comes out from practice: Practice → Scientific Problem → Scientific Research (statement and checking of hypotheses) → Rational Action. Even so, this is far from being the sole way in which scientific research and action mingle. Ever since theoretical mechanics began, in the eighteenth century, to shape industrial machinery, scientific ideas have been the main motor and technology their beneficiary. Since then, intellectual curiosity has been the source of most, and certainly of all important, scientific problems; technology has often followed in the wake of pure research, with a decreasing time lag between the two.

This is not to debase applied science but to recall how rich its conceptual background is. In applied science a theory is not only the summit of a research cycle and a guide to further research; it is also the basis of a system of rules prescribing the course of optimal practical action. On the other hand, in the arts and crafts theories are either absent or instruments of action alone. In past epochs a man was regarded as practical if, in acting, he paid little or no attention to theory or if he relied on worn-out theories and common knowledge. Nowadays a practical man is one who acts in obedience to decisions taken in the light of the best technological knowledge – not pure scientific knowledge, because this is mostly remote from or even irrelevant to practice. And such a technological knowledge, made up of theories, grounded rules, and data, is in turn an outcome of the application of the method of science to practical problems.

Since technology is as theory laden as pure science, and since this either is overlooked or explicitly denied by most philosophers, we must take a closer look at technological theories and their application.

II. TECHNOLOGICAL THEORIES: SUBSTANTIVE AND OPERATIVE

A theory may have a bearing on action either because it provides knowledge regarding the objects of action, for example, machines, or because it is concerned with action itself, for example, with the decisions that precede and steer the manufacture or use of machines. A theory of flight is of the former kind, whereas a theory concerning the optimal decisions regarding the distribution of aircraft over a territory is of the latter kind. Both are *technological theories* but, whereas the theories of the first kind are *substantive*, those of the second kind are, in a sense, *operative*. Substantive technological theories are essentially applications, to nearly real situations, of scientific theories; thus, a theory of flight is essentially an application of fluid dynamics. Operative technological theories, on the other hand, from the start are concerned with the operations of men and man-machine complexes in nearly real situations; thus, a theory of airways management does not deal with planes but with certain operations of the personnel. Substantive technological theories are always preceded by scientific theories, whereas operative theories are born in applied research and may have little if anything to do with substantive theories – this being why mathematicians and logicians with no previous scientific training can make important contributions to them. A few examples will make the substantive-operative distinction clearer.

The relativistic theory of gravitation might be applied to the design of generators of antigravity fields (i.e., local fields counteracting the terrestrial gravitational field), which in turn might be used to facilitate the launching of spaceships. But, of course, relativity theory is not particularly concerned with either field generators or astronautics; it just provides some of the knowledge relevant to the design and manufacture of antigravity generators. Paleontology is used by the applied geologist engaged in oil prospecting, and the latter's findings are a basis for making decisions concerning drillings; but neither paleontology nor geology is particularly concerned with the oil industry. Psychology can be used by the industrial psychologist in the interests of production, but it is not basically concerned with production. All three are examples of the application of scientific (or semiscientific, as the case may be) theories to problems that arise in action.

On the other hand the theories of value, decision, games, and operations research deal directly with valuation, decision-making, planning, and doing; they even may be applied to scientific research regarded as a kind of action, with the optimistic hope of optimizing its output. (These theories could not tell how to replace talent but how best to exploit it.) These are operative theories, and they make little if any use of the substantive knowledge provided by the physical, biological, or social sciences: ordinary knowledge, special but non-scientific knowledge (of, e.g., inventory practices), and formal science are usually sufficient for them. Just think of strategical kinematics applied to combat or of queuing models: they are not applications of pure scientific theories but theories on their own.

What these operative or non-substantive theories employ is not substantive scientific knowledge but the *method* of science. They may be regarded, in fact, as scientific theories concerning actions, in short, as theories of action. These theories are technological in respect of aim, which is practical rather than cognitive, but apart from this they do not differ markedly from the theories of science. In fact, every good operative theory will have at least the following traits characteristic of scientific theories: (1) they do not refer directly to chunks of reality but to more or less idealized models of them (e.g., entirely rational and perfectly informed contenders or continuous demands and deliveries); (2) as a consequence they employ theoretical concepts (e.g., 'probability'); (3) they can absorb empirical information and in turn can enrich experience by providing predictions or retrodictions; and (4) consequently they are empirically testable, though not as toughly as scientific theories.

Looked at from a practical angle, technological theories are richer than the theories of science in that, far from being limited to accounting for what may or does, did or will *happen* regardless of what the decision-maker does, they are concerned with finding out *what ought to be done* in order to bring about, prevent, or just change the pace of events or their course in a preassigned way. In a conceptual sense, the theories of technology are definitely poorer than those of pure science; they are invariably *less deep*, and this because the practical man, for whom they are intended, is chiefly interested in net effects that occur and are controllable on the human scale; he wants to know how things within *his* reach can be made to work *for him*, rather than how things of any kind really are. Thus, the electronics expert need not worry about the difficulties that plague the quantum electron theories; and the researcher in utility theory, who is concerned with comparing people's preferences, need not burrow into the origins of preference patterns – a problem for psychologists.

Consequently, whenever possible the applied researcher will attempt to schematize his system as a *black box*; he will deal preferably with external variables (input and output), will regard all others as at best handy intervening variables with no ontological import, and will ignore the adjoining levels. This is why his oversimplifications and mistakes are not more often harmful – because his hypotheses are superficial. (Only the exportation of this externalist approach to science may be harmful.) Occasionally, though, the technologist will be forced to take up a deeper, representational viewpoint. Thus, the molecular engineer who designs new materials to order, that is, substances with prescribed macroproperties, will have to use certain fragments of atomic and molecular theory. But he will neglect all those microproperties that do not show up appreciably at the macroscopic level; after all, he uses atomic and molecular theories as tools – which has misled some philosophers into thinking that scientific theories are *nothing but* tools.

The conceptual impoverishment undergone by scientific theory when used as a means for practical ends can be frightful. Thus, an applied physicist engaged in designing an optical instrument will use almost only ray optics, that is, essentially what was known about light toward the middle of the seventeenth century. He will take wave optics into account for the explanation in outline, not in detail, of some effects, mostly undesirable, such as the appearance of colors near the edge of a lens; but he will seldom, if ever, apply any of the various wave theories of light to the computation of such effects. He can afford to ignore these theories in most of his professional practice because of two reasons. First, the chief traits of the optical facts relevant to the manufacture of optical instruments are adequately accounted for by ray optics; those few facts that are not so explainable require only the hypotheses (but not the whole theory) that light is made up of waves and that these waves can superpose. Second, it is extremely difficult to solve the wave equations of the deeper theories save in elementary cases, which are mostly of a purely academic interest (i.e., which serve essentially the purpose of illustrating and testing the theory). Just think of the enterprise of solving a wave equation with time-dependent boundary conditions such as those representing the moving shutter of a camera. Wave optics is scientifically important because it is nearly true; but for most present-day technology it is less important than ray optics, and its detailed application to practical problems in optical industry would be quixotic. The same argument can be carried over to the rest of pure science in relation to technology. The moral is that, if scientific research had sheepishly applied itself to the immediate needs of production, we would have no pure science, hence no applied science either.

III. DOES PRACTICE VALIDATE THEORY?

A theory, if true, can be employed successfully in applied research (technological investigation) and in practice itself – as long as the theory is relevant to either. (Fundamental theories are not so applicable because they deal with problems much too remote from practical problems. Just think of applying the quantum theory of scattering to car collisions.) But the converse is not true, that is, the practical success or failure of a scientific theory is no objective index of its truth value. In fact, a theory can be both successful and false; or, conversely, it can be a practical failure and nearly true. The efficiency of a false theory may be due to either of the following reasons. First, a theory may contain just a grain of truth, and this grain alone is employed in the theory's applications. In fact, a theory is a system of hypotheses, and it is enough for a few of them to be true or nearly so in order to be able to entail adequate consequences if the false ingredients are not used in the deduction or if they are practically innocuous. Thus, it is possible to manufacture excellent steel by combining magical exorcisms with the operations prescribed by the craft – as was done until the beginning of the nineteenth century. And it is possible to improve the condition of neurotics by means of shamanism, psychoanalysis, and other practices as long as effective means, such as suggestion, conditioning, tranquilizers, and above all time are combined with them.

A second reason for the possible practical success of a false theory may be that the accuracy requirements in applied science and in practice are far below those prevailing in pure research, so that a rough and simple theory supplying quick correct estimates of orders of magnitude very often will suffice in practice. Safety coefficients will mask the finer details predicted by an accurate and deep theory anyway, and such coefficients are characteristic of technological theory because this must adapt itself to conditions that can vary within ample bounds. Think of the variable loads a bridge can be subjected to or of the varying individuals that may consume a drug. The engineer and the physician are interested in safe and wide intervals centered in typical values rather than in exact values. A greater accuracy would be pointless since it is not a question of testing. Moreover, such a greater accuracy could be confusing because it would complicate things to such an extent that the target – on which action is to be focused – would be lost in a mass of detail. Extreme accuracy, a goal of scientific research, not only is pointless or even encumbering in practice in most cases but can be an obstacle to research itself in its early stages. For the two reasons given above – use of only a part of the premises and low accuracy requirements – infinitely many possible rival theories can yield 'practically the same results'. The technologist, and particularly the technician, are justified in preferring the simplest of them: after all, they are interested primarily in efficiency rather than in truth, in getting things done rather than in gaining a deep understanding of them. For the same reason, deep and accurate theories may be impractical; to use them would be like killing bugs with nuclear bombs. It would be as preposterous – though not nearly so dangerous – as advocating simplicity and efficiency in pure science.

A third reason why most fundamental scientific theories are of no practical avail is not related to the handiness and sturdiness required by practice but has a deep ontological root. The practical transactions of man occur mostly on his own level; and this level, like others, is rooted to the lower levels but enjoys a certain autonomy with respect to them, in the sense that not every change occurring in the lower levels has appreciable effects on the higher ones. This is what enables us to deal with most things on their own level, resorting at most to the immediately adjacent levels. In short, levels are to some extent stable: there is a certain amount of play between level and level, and this is a root of both chance (randomness due to independence) and freedom (self-motion in certain respects). One-level theories will suffice, therefore, for many practical purposes. It is only when a knowledge of the relations among the various levels is required in order to implement a 'remote-control' treatment, that many-level theories must be tried. The most exciting achievements in this respect are those of psychochemistry, the goal of which is, precisely, the control of behavior by manipulating variables in the underlying biochemical level.

A fourth reason for the irrelevance of practice to the validation of theories – even to operative theories dealing with action – is that, in real situations, the relevant variables are seldom adequately known and precisely controlled. Real situations are much too complex for this, and effective action is much too strongly urged to permit a detailed study – a study that would begin by isolating variables and tying some of them into a theoretical model. The desideratum being maximal efficiency, and not at all truth, a number of practical measures will usually be attempted at the same time: the strategist will counsel the simultaneous use of weapons of several kinds, the physician will prescribe a number of supposedly concurrent treatments, and the politician may combine promises and threats. If the outcome is satisfactory, how will the practitioner know which of the rules was efficient, hence which of the underlying hypotheses was true? If unsatisfactory, how will he be able to weed out the inefficient rules and the false underlying hypotheses?

A careful discrimination and control of the relevant variables and a critical evaluation of the hypotheses concerning the relations amoung such variables is not done while killling, curing, or persuading people, not even while making things, but in leisurely, planned, and critically alert scientific theorizing and experimentation. Only while theorizing or experimenting do we *discriminate* among variables and *weigh* their relative importance, do we *control* them either by manipulation or by measurement, and do we *check* our hypotheses and inferences. This is why factual theories, whether scientific or technological, substantive or operative, are empirically tested in the laboratory and not in the battlefield, the consulting office, or the market place. ('Laboratory' is understood here, in a wide sense, to include any situation which, like the military maneuver, permits a reasonable control of the relevant variables.) This is, also, why the efficiency of the rules employed in the factory, the hospital, or the social institution, can be determined only in artificially controlled circumstances.

In short, practice has no validating force; pure and applied research alone can estimate the truth value of theories and the efficiency of technological rules. What the technician and the practical man do, by contrast to the scientist, is not to *test* theories but to *use* them with non-cognitive aims. (The practitioner does not even test *things*, such as tools or drugs, save in extreme cases: he just uses them, and their properties and their efficiency again must be determined in the laboratory by the applied scientist.) The doctrine that practice is the touchstone of theory relies on a misunderstanding of both practice and theory, on a confusion between practice and experiment and an associated confusion between rule and theory. The question 'Does it work?' – pertinent as it is with regard to things and rules – is impertinent in respect of theories.

Yet it might be argued that a man who knows how to do something is thereby showing that he knows that something. Let us consider the three possible versions of this idea. The first can be summed up in the schema 'If x knows how to do (or make) y, then x knows y.' To ruin this thesis it is enough to recall that, for nearly one million years, man has known how to make children without having the remotest idea about the reproduction process. The second thesis is the converse conditional, namely, 'If x knows y, then x knows how to do (or make) y.' Counterexamples: we know something about stars, yet we are unable to make them, and we know part of the past, but we cannot even spoil it. The two conditionals being false, the biconditional 'x knows y if and only if x knows how to do (or make) y' is false, too. In short, it is false that knowledge is identical with knowing how to do, or know-how. What is true is rather this: knowledge considerably *improves* the chances of correct doing, and doing *may* lead to knowing more (now that we have learned that knowledge pays), not because action is knowledge, but because, in inquisitive minds, action may trigger questioning.

It is only by distinguishing scientific knowledge from instrumental knowledge, or know-how, that we can hope to account for the coexistence of practical knowledge with theoretical ignorance and the coexistence of theoretical knowledge with practical ignorance. Were it not for this the following combinations hardly would have occurred in history: (1) science without the corresponding technology (e.g., Greek physics); (2) arts and crafts without an underlying science (e.g., Roman engineering and contemporary intelligence testing). The distinction must be kept, also, in order to explain the cross-fertilizations of science, technology, and the arts and crafts, as well as to explain the gradual character of the cognitive process. If, in order to exhaust the knowledge of a thing, it were sufficient to produce or reproduce it, then certain technological achievements would put an end to the respective chapters of applied research: the production of synthetic rubber, plastic materials and synthetic fibres would exhaust polymer chemistry; the experimental induction of cancer should have stopped cancer research; and the experimental production of neuroses and psychoses should have brought psychiatry to a halt. As a matter of fact, we continue doing many things without understanding how, and we know many processes (such as the fusion of helium out of hydrogen) which we are not yet able to control for useful purposes (partly because we are too eager to attain the goal without a further development of the means). At the same time it is true that the barriers between scientific and practical knowledge, pure and applied research, are melting. But this does not eliminate their differences, and the process is but the outcome of an increasingly scientific approach to practical problems, that is, of a diffusion of the scientific method.

The identification of knowledge and practice stems not only from a failure to analyze either but also from a legitimate wish to avoid the two extremes of speculative theory and blind action. But the testability of theories and the possibility of improving the rationality of action are not best defended by blurring the differences between theorizing and doing, or by asserting that action is the test of theory, because both theses are false and no program is defensible if it rests on plain falsity. The interaction between theory and practice and the integration of the arts and crafts with technology and science are not achieved by proclaiming their unity but by multiplying their contacts and by helping the process whereby the crafts are given a technological basis and technology is entirely converted into applied science. This involves the conversion of the rules of thumb peculiar to the crafts into grounded rules, that is, rules based on laws. Let us approach this problem next.

IV. SCIENTIFIC LAW AND TECHNOLOGICAL RULE

Just as pure science focuses on objective patterns or laws, action-oriented research aims at establishing stable norms of successful human behavior, that is, rules. The study of rules – the grounded rules of applied science – is therefore central to the philosophy of technology.

A rule *prescribes* a course of action; it indicates how one should proceed in order to achieve a predetermined goal. More explicitly, a rule is an instruction to perform a finite number of acts in a given order and with a given aim. The skeleton of a rule can be symbolized as a string of signs, such as 1-2-3-...-*n*, where every number stands for a corresponding act; the last act, *n*, is the only thing that separates the operator who has executed every operation, save *n*, from the goal. In contrast to law formulas, which say what the shape of possible events is, rules are norms. The field of law is assumed to be the whole of reality, including rule-makers; the field of rule is but mankind; men, not stars, can obey rules and violate them, invent and perfect them. Law statements are descriptive and interpretive, whereas rules are normative. Consequently, while law statements can be more or less true, rules can be only more or less effective.

We may distinguish the following genera of rules: (1) *rules of conduct* (social, moral, and legal rules); (2) *rules of prescientific work* (rules of thumb in the arts and crafts and in production); (3) *rules of sign* (syntactical and semantical rules); (4) *rules of science and technology* (grounded rules of research and action). Rules of conduct make social life possible (and hard). The rules of prescientific work dominate the region of practical knowledge which is not yet under technological control. The rules of sign direct us how to handle symbols – how to generate, transform, and interpret signs. And the rules of science and technology are those norms that summarize the special techniques of research in pure and applied science (e.g., random-sampling techniques) and the special techniques of advanced modern production (e.g., the technique of melting with infrared rays).

Many rules of conduct, work, and sign, are *conventional*, in the sense that they are adopted with no definite reasons and might be exchanged for alternative rules with little or no concomitant change in the desired result. They are not altogether arbitrary, since their formation and adoption should be explainable in terms of psychological and sociological laws, but they are not necessary either; the differences among cultures are largely differences among systems of rules of that kind. We are not interested in such groundless or conventional rules but rather in founded rules, that is, in norms satisfying the following *definition*: A rule is *grounded* if and only if it is based on a set of law formulas capable of accounting for its effectiveness. The rule that commands taking off the hat when greeting a lady is groundless in the sense that it is based on no scientific law but is conventionally adopted. On the other hand, the rule that commands greasing cars periodically is based on the law that lubricators decrease the wearing out of parts by friction; this is neither a convention nor a rule of thumb like those of cooking and politicking – it is a well-grounded rule.

To decide that a rule is effective it is necessary, though insufficient, to show that it has been successful in a high percentage of cases. But these cases might be just coincidences, such as those that may have consecrated the magic rituals that accompanied the huntings of primitive man. Before adopting an empirically effective rule we ought to know *why* it is effective; we ought to take it apart and reach an understanding of its *modus operandi*. This requirement of rule foundation marks the transition between the prescientific arts and crafts and contemporary technology. Now, the sole valid foundation of a rule is a system of law formulas, because these alone can be expected to correctly explain facts, for example, the fact that a given rule works. This is not to say that the effectiveness of a rule depends on whether it is founded or groundless but only that, in order to be able to *judge* whether a rule has any chance of being effective, as well as in order to *improve* the rule and eventually *replace* it by a more effective one, we must disclose the underlying law statements, if any. We may take a step ahead and claim that the blind application of rules of thumb has never paid in the long run; the best policy is, first, to try to ground our rules and, second, to try to transform some law formulas into effective technological rules. The birth and development of modern technology is the result of these two movements.

But it is easier to preach the foundation of rules than to say exactly what the foundation of rules consists in. Let us try to make an inroad into this unexplored territory – the core of the philosophy of technology. As usual when approaching a new subject, it will be convenient to begin by analyzing a typical case. Take the law statement 'Magnetism disappears above the Curie temperature (770 °C for iron)'. For purposes of analysis it will be convenient to restate our law as an explicit conditional: 'If the temperature of a magnetized body exceeds its Curie point, then it becomes demagnetized'. (This is, of course, an oversimplification, as every other ordinary-language rendering of a scientific law: the Curie point is not the temperature at which all magnetism disappears but, rather, the point of conversion of ferromagnetism into paramagnetism, or conversely. But this is a refinement irrelevant to most technological purposes.) Our nomological statement provides the basis for the nomopragmatic statement 'If a magnetized body is heated above its Curie point, then it is demagnetized'. (The pragmatic predicate is, of course, 'is heated'.) This nomopragmatic statement is, in turn, the ground for two different rules, namely, R1: 'In order to demagnetize a body heat it above its Curie point', and R2: 'To prevent demagnetizing a body do not heat it above its Curie point'. Both rules have the same foundation, that is, the same underlying nomopragmatic statement, which in turn is supported by a law statement assumed to represent an objective pattern. Moreover, the two rules are equiefficient, though not under the same circumstances (changed goals, changed means).

Notice, first, that unlike a law statement a rule is neither true nor false; as a compensation it can be effective or ineffective. Second, a law is consistent with more than one rule. Third, the truth of a law statement does not insure the efficiency of the associated rules; in fact, the former refers to idealized situations which are not met with in practice. Fourth, whereas given a law we may try out the corresponding rules, given a rule we are unable to trace the laws presupposed by it; in fact, a rule of the form 'In order to attain the goal *G* employ the means *M*' is consistent with the laws 'If *M*, then *G*', '*M* and *G*', '*M* or *G*', and infinitely many others.

The above has important consequences for the methodology of rules and the interrelations between pure and applied science. We see there is no single road from practice to knowlegde, from success to truth; success warrants no inference from rule to law but poses the problem of explaining the apparent efficiency of the rule. In other words, the roads from success to truth are infinitely many and consequently theoretically useless or nearly so, that is, no bunch of effective rules suggests a true theory. On the other hand, the roads from truth to success are limited in number, hence feasible. This is one of the reasons why practical success, whether of a medical treatment or of a government measure, is not a truth criterion for the underlying hypotheses. This is also why technology – in contrast to the prescientific arts and crafts – does not start with rules and end up with theories but proceeds the other way around. This is, in brief, why technology is applied science whereas science is not purified technology.

Scientists and technologists work out rules on the basis of theories containing law statements and auxiliary assumptions, and technicians apply such rules jointly with groundless (prescientific) rules. In either case, specific hypotheses accompany the application of rules, namely, hypotheses to the effect that the case under consideration is one where the rule is in point because such and such variables – related by the rule – are in fact present. In science such hypotheses can be tested; this is true of both pure and applied research. But in the practice of technology there may not be time to test them in any way other than by applying the rules around which such hypotheses cluster – and this is a poor test indeed, because the failure may be blamed either on the hypotheses or on the rule or on the uncertain conditions of application.

V. SCIENTIFIC PREDICTION AND TECHNOLOGICAL FORECAST

For technology knowledge is chiefly a means to be applied to the achievement of certain practical ends. The goal of technology is successful action rather than pure knowledge, and accordingly the whole attitude of the technologist while applying his technological knowledge is active in the sense that, far from being an inquisitive onlooker or a diligent burrower, he is an active participant in events. This difference of attitude between the technologist in action and the researcher – whether pure or applied – introduces certain differences between technological forecast and scientific prediction.

In the first place, whereas scientific prediction says what will or may happen if certain circumstances obtain, technological forecast suggests how to influence circumstances so that certain events may be brought about, or prevented, that would not normally happen; it is one thing to predict the trajectory of a comet, quite another to plan and foresee the orbit of an artificial satellite. The latter presupposes a choice among possible goals, and such a choice presupposes a certain forecasting of possibilities and their evaluation in the light of a set of desiderata. In fact, the technologist will make his forecast on his (or his employer's) estimate of what the future *should* be like if certain desiderata are to be fulfilled; contrary to the pure scientist, the technologist is hardly interested in what would happen anyway; and what for the scientist is just the final state of a process becomes for the technologist a valuable (or disvaluable) end to be achieved (or to be avoided). A typical scientific prediction has the form 'If x occurs at time t, then y will occur at time t' with probability p'. By contrast, a typical technological forecast is of the form 'If y is to be achieved at time t' with probability p, then x should be done at time t'. Given the goal, the technologist indicates the adequate means, and his forecast states a means-end relationship rather than a relation between an initial state and a final state. Furthermore, such means are implemented by a specified set of actions, among them the technologist's own actions.

This leads us to a second peculiarity of technological forecast: whereas the scientist's success depends on his ability to separate his object from himself (particularly so when his object happens to be a psychological subject) – that is, on his capacity of detachment – the technologist's ability consists in placing himself within the system concerned – at the head of it. This does not involve *subjectivity*, since after all the technologist draws on the objective knowledge provided by science; but it does involve partially, a *parti pris* unknown to the pure researcher. The engineer is part of a man-machine complex, the industrial psychologist is part of an organization, and both are bound to devise and implement the optimal means for achieving desiderata which are not usually chosen by themselves; they are decision-makers, not policy-makers.

The forecast of an event or process that is not under our control will not alter the event or process itself. Thus, for example, no matter how accurately an astronomer predicts the collision of two stars, the event will occur in due course. But if an applied geologist can forecast a landslide, then some of its consequences can be prevented. Moreover, by designing and supervising the appropriate defense works the engineer may prevent the landslide itself; he may devise the sequence of actions that will refute the original forecast. Similarly, an industrial concern may forecast sales for the near future on the (shaky) assumption that a given state of the economy, say prosperity, will continue during that lapse. But if this assumption is falsified by a recession, and the enterprise had accumulated a large stock which it must get rid of, then instead of making a new sales forecast (as a pure scientist would be inclined to do), the management will try to *force* the original forecast to come true by increasing advertisement, lowering sale prices, and so on. As in the case of vital processes, a diversity of means will alternatively or jointly be tried to attain a fixed goal. In order to achieve this goal any number of initial hypotheses may have to be sacrified: in the case of the landslide, the assumption that no external forces would interfere with the process and, in the case of the sales, that prosperity would continue. Consequently, whether the initial forecast is *forcefully falsified* (as in the case of the landslide) or *forcefully confirmed* (as in the case of the sales forecast), this fact cannot count as a *test* of the truth of the hypotheses involved; it will count only as an efficiency test of the rules that have been applied. The pure scientist, on the other hand, need not worry about altering the means for achieving a preset goal, because pure science *has* no goals external to it.

Technological forecast, in sum, cannot be used for controlling things or men by changing the course of events perhaps to the point of stopping them altogether, or for forcing the predicted course even if unpredictable events should interfere with it. This is true of the forecasts made in engineering, medicine, economics, applied sociology, political science, and other technologies: the sole formulation of a forecast (prognosis, lax prediction, or prediction proper), if made known to the decision-makers, can be seized upon by them to steer the course of events, thus bringing about results different from those originally forecasted. This change, triggered by the issuance of the forecast, may contribute either to the latter's confirmation (self-fulfilling forecast) or to its refutation (self-defeating forecast). This trait of technological forecast stems from no logical property of it; it is a pattern of social action involving the knowledge of forecasts and consequently is conspicuous in modern society. Therefore, rather than analyzing the logic of causally effective forecast, we should start by distinguishing three levels in it: (1) the conceptual level, on which the prediction p stands; (2) the psychological level – the knowledge of p and the reactions triggered by this knowledge; and (3) the social level – the actions actually performed on the basis of the knowledge of p and in the service of extra-scientific goals. This third level is peculiar to technological forecast.

This feature of technological forecast sets civilized man apart from every other system. A non-predicting system, be it a jukebox or a frog, when fed with information it can digest will process it and convert it into action at some later time. But such a system does not purposely produce most of the information, and it does not issue projections capable of altering its own future behavior. A prediction – a rational man, a team of technologists, or a sufficiently evolved automaton – can behave in an entirely different way. When fed with relevant information I_t at time t, it can process this information with the help of the knowledge (or the instructions) available to it, eventually issuing a prediction P'_t, at a later time t'. This prediction is fed back into the system and compared with the preset goal that controls the whole process (without either causing it or supliying it with energy). If the two are reasonably close, the system takes a decision that eventually leads it to act so as to take advantage of the course of events. If, on the other hand, the prediction differs significantly from the goal, this difference will again trigger the theoretical mechanism, which will elaborate a new strategy: a new prediction, P_t'', will eventually be issued at time t'', a forecast including a reference to the system's own participation in the events. The new prediction is fed back into the system and, if it still disagrees with the goal, a new correction cycle is triggered, and so on until the difference between the prediction and the goal becomes negligible, in which case the system's predicting mechanism comes to rest. Henceforth the system will gather new information regarding the present state of affairs and will act so as to conform to the strategy it has elaborated. This strategy may have required not only new information regarding the external world (including the attitudes and capabilities of the people concerned) but also new hypotheses or even theories which had not been present in the instruction chart originally received by the predictor. If the latter fails to realize it or to obtain and utilize such additional knowledge, his or its action is bound to be ineffective. Moral: the more brains the better.

VI. TECHNOLOGICAL FORECAST AND EXPERT PROGNOSIS

The preceding account of technological forecast is based on the assumption that it relies on some theory, or rather theories, whether substantive or operative. This assumption may be found wanting by anyone knowing that the forecasts issued by experts in medicine, finance, or politics are often successful and yet involve no great deal of theorizing. True, most often *expert prognosis* relies on inductive (empirical) generalizations of the form '*A* and *B* occur jointly with the observed frequency *f*', or even just '*A* and *B* occur jointly in most cases', or 'Usually, whenever *A* then *B*'. The observation that a given individual, say a human subject or an economic state of affairs, has the property *A* is then used to forecast that it has, or will acquire, the property *B*. In daily life such prognoses are all we do, and the same applies to most expert prognoses. Occasionally such prognoses made with either ordinary knowledge or specialized but non-scientific knowledge are more successful than predictions made with full-fledged but false or rough theories; in many fields, however, the frequency of hits is not better than the one obtained by flipping a coin. The point, though, is that expert forecast using no scientific theory is not a scientific activity – just by definition of 'scientific prediction'.

Yet it would be wrong to think that experts make no use of *specialized knowledge* whenever they do not employ scientific theories; they always judge on the basis of some such knowledge. Only, expert knowledge is not always explicit and articulate and, for this reason, it is not readily controllable: it does not learn readily from failures, and it is hard to test. For the progress of science, the failure of a scientific prediction is by far preferable to the success of an expert prognosis, because the scientific failure can be fed back into the theory responsible for it, thereby giving us a chance to improve it, whereas in the case of expert knowledge there is no theory to feed the failure into. It is only for immediate practical purposes that expert prognoses made with shallow but well-confirmed generalizations are preferable to risky scientific predictions.

Another difference between expert prognosis and technological forecast proper would seem to be this: the former relies more heavily on *intuition* than does scientific prediction. Yet the difference is one of degree rather than of kind. Diagnosis and forecast, whether in pure science, in applied science, or in the arts and crafts, involve intuitions of a number of kinds: the quick identification of a thing, event, or sign; the clear but not necessarily deep grasp of the meaning and/or the mutual relations of a set of signs (text, table, diagram, etc.); the ability to interpret symbols; the ability to form space models; skill in realizing analogies; creative imagination; catalytic inference, that is, quick passage from some premises to other formulas by skipping intermediate steps; power of synthesis or synoptic grasp; common sense (or rather controlled craziness), and sound judgment. These abilities intertwine with specialized knowledge, whether scientific or not, and are reinforced with practice. Without them theories could neither be invented nor applied – but, of course, they are not suprarational powers. Intuition is all right as long as it is controlled by reason and experiment; only the replacement of theorizing and experimenting by intuition must be feared.

A related danger is that of *pseudoscientific projection tools*, so common in applied psychology and sociology. A number of techniques have been devised to forecast the performance of personnel, students, and even psychologists themselves. A few tests, the objective ones, are somewhat reliable; this holds for intelligence and skill tests. But most tests, particularly the subjective ones (the 'global evaluation' of personality by means of interviews, the thematic apperception tests, the Rorschach, etc.) are in the best of cases inefficient and in the worst of cases misleading. When they have been subjected to the test of prediction – that is, when their results have been checked with the actual performance of the subjects – they have failed. The failure of most individual psychological tests, and particularly of the subjective ones, is not a failure of psychological testing in general; what is responsible for such failures is either the total absence or the falsity of the underlying psychological theories. Testing for human abilities without first establishing *laws* relating objective indexes of abilities or personality traits is as thoughtless as asking a tribesman to test an aircraft. As long as no theoretical foundations of psychological tests are secured, their employment as predictive instruments is not better than crystal gazing or coin-flipping: they are practically inefficient and, even if they succeeded, they would not contribute to psychological theory, because they are unrelated to theory. The limited success of psychological testing has led many to despair of the possibility of finding a scientific approach to human behavior, but the right inference is that such an attempt has been tried only after a large number of alleged tests invaded the market. What is wrong with most of 'applied' (educational, industrial, etc.) psychology is that it does *not* consist in the application of scientific psychology at all. The moral is that practical wants – such as personnel training and selection – should not be allowed to force the construction of 'technologies' without an underlying science.

Technological forecast should be maximally reliable. This condition excludes from technological practice – not, however, from technological research – insufficiently tested theories. In other words, technology will ultimately prefer the old theory that has rendered distinguished service in a limited domain and with a known inaccuracy to the bold new theory that promises unheard of forecasts but is probably more complex and therefore partly less well tested. It would be irresponsible for an expert to apply a new idea in practice without having tested it under controlled conditions. (Yet this is still done in pharmacy: recall the affair of the mutagenic drugs in the early 1960's.) Practice, and even technology, is bound to be more conservative than science. Consequently, the effects of a close association of pure research with applied research, and of the latter with production, are not all of them beneficial; while it is true that technology challenges science with new problems and supplies it with new equipment for data-gathering and data-processing, it is no less true that technology, by its very insistence on reliability, standardization (routinization), and speed, at the expense of depth, range, accuracy, and serendipity, can slow down the advancement of science.

VII. OTHER PROBLEMS

We have looked into a few problems of the philosophy of technology Many other challenging problems have been left out, for example, the logic of technological rules; the test of technological theories; the patterns of technological invention; the reason textile, aircraft, and other industries are still largely based on crafts; and the power of technology to bring together previously separate fields (cases of cybernetics, nuclear engineering, computer science, space science, and bioengineering). These and many other problems are waiting to be discovered and worked out by philosophers attentive to their own times. Why should the waiting time be so long?

Montreal

NOTE

* First published in *Technology and Culture* 7 (1966), 329–347.

J. AGASSI

THE CONFUSION BETWEEN SCIENCE AND TECHNOLOGY IN THE STANDARD PHILOSOPHIES OF SCIENCE*

The distinction between pure and applied science seems too trivial to draw, since applied science, as the name implies, aims at practical ends, whereas pure science does not. There is an overlap, to be sure, which is known as fundamental research and which is pure science in the short run but applied in the long run; that is to say, fundamental research is the search for certain laws of nature with an eye to using these laws. Still, this overlap shows that though the distinction is not exclusive it is clear enough. The distinction between applied science and technology is a different matter altogether. All philosophers of science equate them, whereas it is clear that technology includes, at the very least, applied science, invention, implementation of the results of both applied science and invention, and the maintenance of the existing apparatus, especially in the face of unexpected changes, disasters, and so forth. The distinction between applied science and invention, to my knowledge, was made by only one writer, the most important writer on technology, perhaps; I am referring to H. S. Hatfield and his *The Inventor and His World.* Hatfield does not draw the distinction explicitly, but he uses it clearly and systematically enough. Applied science, according to his view, is an exercise in deduction, whereas invention is finding a needle in a haystack.

My own concern with all this comes from my studies in the theory of confirmation. Contrary to most, if not all, writers in the field, I hold that confirmation plays no significant role in science, pure or applied.[1] The contrary impression seems to me to stem from the fact that both invention and the implementation of novelties – from applied science or from invention – require confirmation. The standards of confirmation are legal, and they are set by patent offices in the case of invention and by institutions in charge of public safety and of commercial practices in the case of implementation. The excessive demand for confirmation is the tool by which the complacent postpones the implementation of novelties.

I shall return to the distinction between applied science and invention later on and also touch on the other points of technology. Here let me say that the literature in the philosophy of science is oblivious of all this, usually confusing pure science with applied science and both with technology. The inductivist philosophers of science insist that pure science concerns probable beliefs and that technology concerns decisions, both relating to the same probable hypotheses. They confuse belief and decision. Instrumentalist philosophers do not believe in the probability of pure science but in its usefulness: not only do they equate pure with applied science and applied science with technology, but they even consider all three as identical. Both leading groups of philosophers of science appeal to the fact that science and technology are so very enormously successful. To go to the root of their confusion[2] let us examine views on success. We shall later be able to show how different scientific success is from practical success.

I

Living in a robust progressive society, we are surrounded by the tendency to evaluate people by the measure of their success, more specifically by the measure of their success in attaining high economic and social standing, which is usually known – rather inadequately, but let this ride – as material success. Obviously, success philosophy is rather vulgar, and material success philosophy particularly vulgar and even stupid. I have no intention of speaking against success, for whithout some success there is no action; nor do I wish to speak against material success.

But to value success is not to say that success is all that matters, or that material success is all that we value, with the exception of saints, perhaps. Many works of fiction comprise plausible counter-examples to the identification of success with material success, usually by describing convincingly heroes who are wretched though materially successful (e.g., Bergman's *Wild Strawberries*). A less convincing yet true counter-example is Thomas Alva Edison. He had been very poor and miserable throughout childhood and adolescence, and he consequently loved both money and fame – even to an exaggeration; and he was no saint, being really a difficult man and not always honest, and being prejudiced against intellectuals, women, and Jews; yet he always insisted in words and subborn deeds that the challenge of invention was to him in itself much more of a remuneration than all the benefit it accrued him, financially or socially. He even brought upon himself material ruin so as to put pressure upon himself to work hard at his inventions in order to make money which he then quickly disposed of again, and so on, until he was too old to continue.

This example shows how complex and unconvincing may be the case of any person's personal success. And so let us ignore personal success from now on, material or otherwise. When we speak of success impersonally, we speak of the successful execution of a task regardless of its personal significance to the one who carries it out. Can we evaluate a person by the measure of the success of his performance? Can we measure the man by the measure of his achievement? Obviously, this, too, is rather vulgar. The latest biography of Edison, for instance, by Matthew Josephson, has been written partly, if not mainly, in order to break away from the vulgar success philosophy which permeated previous biographies, thus making them unpalatable to the more sophisticated.[3] At the very least, we all must agree, if we do measure people by their success, then we should take into account the fact that success or its absence depends partly on luck, good or bad; we must adjust the evaluation of the person, then, by considering the factors which were beyond his control yet affected the outcome of his activities. Judging people by their luck is poor judgment. How they can stand up against bad luck and how they can make use of their good luck contribute to their personal makeup much more than how lucky they were.

II

My colleagues in the field of the philosophy of science love to speak of the greatness of science as being identical with the success, the great achievements, of science. Consequently, when we speak of the adventure of science, we are hypocritical: we all admire Captain Scott, since though he failed he was very brave; but Joseph Priestley was seldom accorded similar consideration. Not Priestley, but his opponent, Lavoisier, is admired – and not for his sense of adventure either, but for his alleged success.[4] When you stop to think about it, you may find that success is rather bewildering and calling for much explanation and re-examination.

Somehow the assuredness and faith some of us display in the success of science seem rather smug. This impression is usually dismissed by the claim that the faith in science is amply justified. But when the faith in science is justified, science may become much less of an adventure and, hence, not much of an achievement. The justification of science makes it stupid rather than prudent not to apply it. Moreover, the justification is a principle of induction, and the principle of induction is usually justified by our successful reliance on science in practical affairs. This is circular, but never mind; the principle of induction may indeed be a method for success; it may be the golden goose which lays golden eggs regularly, or a computer which makes its followers rich. If so, then we may compliment the scientists and their followers no more than we may compliment the owner of a golden goose or of a computer. If there exists an inductive algorism which can be fed into a computer and which is most successful, then science must be least exciting and devoid of all adventure. This ideal of science, especially of applied science, as based on a precise algorism, can be found already in Laplace's *Philosophical Essay on Probabilities*, where he expresses his hope to see even judges on the bench replaced by computers.[5] Perhaps the judges of his day were so inhuman that replacing them by machines would have been progress from the negative to the zero degree of humanity; still, the very thought is chilling.

Those, however, who think that viewing science as based on an algorism debases science usually stress that in science we need both luck and intuition. Both words, 'luck' and 'intuition', are indicative of our ignorance, perhaps of our essential ignorance. For, obviously, if this ignorance were temporary, science based on algorism would be attainable. Though they are both mysterious, luck and intuition seem to be opposite poles: whereas we admire those who contribute to scientific progress when we view them as inspired, we somewhat deprecate them when we say they were lucky. Understandably, Pasteur and Edison alike felt deprecated when their success was viewed as due mainly to luck. They both repeated Lagrange's mystical formula about luck coming only to the prepared mind. One might push the argument and say that even having a prepared mind depends on luck. Indeed, both Oersted and Einstein humbly viewed their own inspiration and talent as mere luck. And yet pushing the argument so far is scholasticism: if a typical scientist were lucky enough to have one of two wishes granted, these being for talent and for material success, he would doubtlessly grasp his luck and wish for talent, as King Solomon is alleged to have done, hoping to have the chance to use it. Hence, it is neither talent nor luck but, rather, adventure, namely, the bold use of talent and the grasping of luck when it comes one's way, that we wish. It is all too easy *not* to use either talent or luck by the mere lack of a sufficient degree of courage and perseverance. As Edison said so aptly, "It has been so in all my inventions. The first step is an intuition – and it comes with a burst; *then*, difficulties arise .... I have the right principle, and am on the right track, but time, hard work, and some good luck are necessary too."[6] And again, "The trouble with other inventors is that they try a few things and quit. I *never quit* until I get what I want."[7] Of course, Edison was very lucky to be able to work so hard, so boldly, so cleverly: most people can do nothing to alter their predicament. But Edison's luck is so very different from the luck of the born princess or of the person who got rich at his first gamble; it was not the luck that brought success but the luck that made him able to strive for success.

III

Query: can we admire science as the product of the bold use of the investigator's imagination and of his good fortunes, and yet justify our trust in science? If we justify our trust by a criterion or a formula, the criterion or formula may turn into a Laplacian algorism which renders science mechanical, but it we have no criterion or formula, we may have no justification. Thus, in his *Logical Foundations of Probability*, Carnap tries to differ from Laplace but fails. He indorses the view – which he attributes to Einstein and Popper, but which might be more justly attributed to Brewster and Whewell, if not to Galileo and Kant – of the imagination as an essential ingredient in science.[8] And yet he provides the Laplacian formula for deciding which hypothesis to believe in and to act upon, and this formula provides the Laplacian algorism of projecting the past into the immediate future and thus generating mechanically the best possible hypothesis.[9]

One cannot argue cogently against the creation of any working algorism, that is to say, against any systematization. Descartes systematized geometry and enabled people to prove theorems mechanically rather than cleverly; solid-state physics is now systematizing one of the most known fields of clever invention, namely, metallurgy; and yet no one is any the worse for these systematizations. When Abraham Wald systematized some decision-procedures, his ideas were so powerful that they were kept for a time as war secrets.[10] Any partial systematization, such as Descartes's or Wald's, merely covers some ground and sends the adventurer further afield in search of new frontiers. But total systematization excludes all adventure. In any field, any algorism is welcome; yet, were an algorism universal in that field, creative thinking in that field would be redundant altogether, once and for all. If the field in question is the generation of ideas in general, algorism in it is the end of creative thinking in all science and technology.

But it is possible to invent a criterion or formula for justifying our faith in and reliance on a scientific theory without taking away the spirit of adventure from science. A formula for trusting science which is not a means of generating mechanically the theory to be trusted is possible, though Carnap, at least, has tried to produce it and failed. Such a formula was invented by William Whewell[11] and reinvented with improvement and modification by my teacher, Karl Popper;[12] my dissent from his philosophy is rooted in my rejection of the formula, but I must acknowledge its superiority. Let me present it first and discuss the question whether it is a measure of success later.

It runs as follows. A belief in a theory is justified to the degree to which it was corroborated by experience; that degree depends on the explanatory power of the theory in question, of the degree of testability of that theory, and of the degree of severity of the test which it has thus far passed successfully.

Let us not examine this formula in detail, so as to avoid a rather academic exercise. Let us first try to get the spirit, the general feel and approach behind it, and see how acceptable that is. Up to a point, I think, it is, but only up to a point. To begin with, let us consider the positive general characteristics of this formula. The formula takes full account of the challenge of research. Thus, it is in full accord with the above quotations from Edison, as well as with the following one: "I would construct a theory and work on its lines until I found it untenable, then it would be discarded and another theory evolved. This was the only possible way for me to work out the problem."[13]

Both Whewell and Popper stress the need for problems, for inspired solutions, for the usefulness of criticism in the development of new solutions, and for the success of theories which stood up to criticism and thus proved their mettle. The chief difference between Whewell and Popper, it is well known, is that Whewell believed in the finality of such success whereas Popper believes in its tentativity. Also, one need hardly say, in this division no one sides with Whewell today, even though most philosophers of science accept neither view. The majority indorse a view which is nearer to Whewell's than to Popper's in that they replace Whewell's finality not with Popper's tentativity but with probability. This notion is either meaningless or else it is a theory of probability in the classical sense, and thus it yields a Laplacian algorism.[14]

IV

So now we have a formula – the Whewell-Popper formula – which perhaps justifies our faith in science, and if so, without doing injustice to intuition and luck. Let me now show you that to judge the greatness of science by its success, when success is measured by the Whewell-Popper formula, does not accord closely with the admiration of science as an active adventure, which I am advocating.

Imagine an Einstein, racking his brain, developing his most inspired ideas, sifting them, elaborating them, deducing from them both explanations and tests; he then retires to his den, full of nervous anticipation. The world notices his work, considers it, concedes that it is possibly admirable, and sits patiently and waits. Then comes an Eddington, translates the test into action by designing all sorts of instruments, by mobilizing funds, by organizing workshops to prepare the instruments, by organizing an expedition, by supervising and participating in the experiment, by calculating the experimental results and comparing them to Einstein's results as calculated from Einstein's theory. All the while Einstein is supposed to be sitting back and waiting, and all the time we are supposed neither to admire nor to dismiss him; and then Eddington may give the green light, and we all burst in admiration for Einstein; or he does not, and we do not.

In such a case we want the scientist to intuit not only a brilliant idea but also certain truths. In such a case the scientist is more of a fortune-teller than I would like to be the case. It was said that Faraday had a nose for the truth; Edison was called the 'wizard of Menlo Park'; maybe all this is true. If so, the truth is not very palatable, at least for those who find it less enjoyable to admire Faraday or Edison because he is a wizard than to admire them as intellectual adventurers.

Have we not gotten into an impasse? That we want success is self-evident: to say that we want success is to say no more than to say that we want to achieve something. If we did not want anything or if we had no expectation of achieving it, we would not act at all. And if we do not want the success of applied science to be a matter of wizarding or uncanny insight, we must have an algorism. So, it seems, if we do not want an algorism, we do want some wizarding! Considering this impasse, we may be more patient with my colleagues who view science as a success-algorism. Michael Polanyi[15] is surprised that most philosophers of science deny that applied science or technology contains an intuitive element when this claim is a standard criterion of all patent offices: a machine that can be created by anybody according to a publicly known algorism will not be granted a patent; the claim for a patent must involve (by legal standards!) a claim for originality. Polanyi exaggerates the wisdom of the legal patenting apparatus (consider the Selden case, and see Edison's 'My 40 Years of Litigations'), yet by and large he is right.[16] Still, we may well understand the philosophers whom he ciritcizes: reluctant to acknowledge the wizardry of the inventor, they decide that invention has an algorism to it, and thus they imply – perhaps unwittingly – that inventors need not be original. They equally imply that scientists need not be original, even though we are not supposed to grant Ph.D.'s for dissertations which we do not judge original – this also by some legal standards. This does not disturb the inductive philosophers who insist on the idea of science as unoriginal.

The idea that science is no magic and hence it is based on some algorism is by no means new, by no means peripheral to the traditions of science, and by no means a merely unintended aspect of inductive philosophy. Its chief corollary from our viewpoint is no news either;

it has been stressed again and again that the measure of a scientist's success is, provided he is a proper scientist, the measure of work he has invested in science and no more. This calls for equal admiration for every hard-working contributor – all contributions will be thankfully received – as was stressed by Bacon, by Boyle, and, more recently, by Malinowski as well as A. P. Usher, leading historian of technology.[17] Bacon also said that inventors should be honored by statues made of different material from gold to wood and in different sizes, depending on the greatness of their inventions. Similarly, the fact that we do not admire in equal measure every hard-working inventor and every anthropologist who has conducted fieldwork is well known. Usher dismisses this inequality as cults of personality and historical fiction; young anthropologists are told that every fieldwork is of equal importance, and young chemists and engineers are told similar stories. Yet most young engineers and anthropologists think that all this is a mere scientific myth. Professors reiterate the myth of equality of all contributions, and students reiterate the myth of great adventures. Who is right? What is the real criterion of success?

The instrumentalist philosophers who identify – with Duhem – pure science and applied science have the idea of simplicity as the criterion of success.[18] Mach viewed simplicity as success of mental economy, which is very different from material success.[19] Material success, no doubt, is often awarded to very complicated theories,[20] and to be honest we need economists to assess such material success. And, of course, we should choose materially successful economists to do this job. Moreover, all economists will agree that the archeology of extinct cultures, for instance, is of hardly any practical value, except as a highbrow means of entertainment, perhaps. Most philosophers of science, however, do not indorse this instrumentalist philosophy, preferring to it some inductive philosophy which justifies their claim that science is useful by extrapolating from past to future material success. They say, "when we make a decision, we base it on induction."

These people just do not understand the essentially adventurous nature of decisions. Neurotic people, it is well known, who do not possess the ability to decide, often say, "I cannot decide because I do not know enough."[21] They obviously deceive themselves: when we know enough we have no problem at all, and so we need not decide at all. When we say we must decide which route to take, we imply that we have no knowledge as to which route is the shorter, or the safer, or the better one by any accepted criteria; alternatively, if we have not accepted a criterion, we may be saying that we do not know which criterion is the correct one. So, one thing is quite clear: we just do not decide by induction.

Let us then examine the sort of decisions which pure scientists, applied scientists, and technologists face; the sort of questions they have to try to answer. This will give us, I hope, some insight to the sort of adventure they may face.

V

One reason why I admire Popper's philosophy of science is just its playing down of success. Roughly, but very seriously, let me put it thus. Everybody admires science for its ability to predict. Meaning, of course, its ability to predict *correctly* or, more precisely, sufficiently correctly to pass as the ability to predict *successfully*. (The criteria of success are vague and invite us to confuse science with technology.) Only Popper, to my knowledge, has stressed that the ability to predict with some degree of precision is what characterizes a scientific theory – regardless of the correctness of the prediction. Popper had predecessors, to be sure; Galileo, Boyle, Faraday, Whewell, Peirce, and Edison spoke favorably of mistaken predictions in science, and doubtless others did too. But they all insisted that finally it was the successful prediction that counted, that the errors were important only as stepping-stones to imminent success which really characterizes science. It was Popper's province to discover that this is not necessarily so.

According to Popper's theory, a scientist needs a good hypothesis. He may work for many years with no success, or with almost no success. Nothing in Planck's researches after 1900 equals his success in 1900, for instance. And so, according to Popper, the adventure of science may lead to no results; there is, in science, such a thing as utter failure. But, according to Popper, once success was achieved in that a good hypothesis has been found, and once the work has been invested to show the goodness of the hypothesis, then even if the hypothesis leads to predictions which are refuted we may value that hypothesis. The goodness of the hypothesis, according to Popper, may be measured by its explanatory power, as well as by its testability, which is the ability to deduce from it predictions which may be checked by experiment and observation, quite independently of whether the predictions are later found to be true or false. Hence, our appreciation of science need not follow the Whewell-Popper formula, since that formula speaks of theories which have yielded only correct predictions thus far.

To my regret, Popper has expressed recently a view quite different from the one I have just ascribed to him. Referring to discussions with myself, he has said a good hypothesis should also be corroborated before it be refuted.[22] He uses the Whewell-Popper formula as a measure of the goodness of a hypothesis or of our appreciation of it, thus demanding from scientists some measure of wizardry, though he does not go as far in applying the formula as others would: he does suggest that once a theory is refuted we then withdraw our belief in it, but he does not suggest that in such a case we also withdraw our appreciation of it. Nevertheless, I wish to ignore this aspect of Popper's writings, and if you do not like it, you may view the ideas I attribute to Popper as his ideas in my modification or in my distortion, as you wish.

My best reason for rejecting the use of the Whewell-Popper formula as a measure of success is that I follow Popper's theory of the process of trial and error in science as something essentially different from trial and error in many other situations, such as more everyday ones. Trial and error is progress by elimination. In small sets of alternatives, elimination by trial and error may insure success. Most everyday situations are like that, and indeed, no one in his senses would view researches in such situations as adventurous. When our television set breaks down, we take it for granted that there is a finite set of possible faults it may have. Usually we test them one by one and finally eliminate the fault; there is no adventure in the search. The classical argument against the theory of induction, as you well know, is that there exist infinitely many possible explanations of known facts. Lord Keynes has acknowledged this criticism and has postulated the principle of limited variety which he has found implicit in earlier inductivist writers.[23] Assuming, quite *ad hoc*, by the way, that only a finite number of hypotheses may explain known facts, he advocates the most unadventurous theory of science.

Consider now the adventurous method of trial and error in everyday life. This is known as looking for a needle in a haystack. The haystack is supposed to be infinite for all practical purposes. It is quite conceivable that one goes on eliminating more and more possibilities about where the needle may be and yet not progress to any appreciable extent toward the correct position of the needle. Looking for a needle in a haystack is adventurous; it demands intuition and luck.

According to Popper, the search for a good hypothesis is looking for a needle in a haystack. But, according to his philosophy, once a hypothesis is found which solves the problem at hand, that is, which explains the facts which puzzle us, and once it is found testable, then progress has been achieved. (He even claims ability to prove this,[24] but let us not press him on this point.) Even if a good hypothesis is refuted, things will never be the same again. The refutation of a previously unrefuted hypothesis is surely a new discovery we partly owe to its inventor. The task after the refutation of the new hypothesis is not the same as before: we now wish to explain, not the same set of facts which the refuted hypothesis has explained, nor the same set of facts plus the refuting facts, but the refuted hypothesis, as a special case and as a first approximation, plus the new fact.[25] What Einstein explained is not all the known astronomical facts but Newtonian astronomy plus the facts it failed to explain, just as Newton explained Kepler's theory and the deviations from it. This is a fundamental methodological point. The only way to evade it is to claim that pure empirical data exist; as you may know, such claims are becoming increasingly difficult to maintain.[26]

The metaphysical theory corresponding to this fundamental methodological point, of explaining not bare facts but previous theories and their refutations, is the theory of the approximation to the truth by levels of explanations. This theory may be false; it does, however, justify Popper's rejection of the claim that scientists must intuit the truth, and his claim that the intuition of a possibly true explanation is enough of a step forward for the time being. If so, then though finding a good solution is like finding a needle in a haystack, the eliminating of that solution as false is not like finding that the needle is not in a small portion of a haystack. Pictorially, scientific theories are not explorations of small portions of a given haystack; they are sieves through which the whole haystack is passed and whose meshes are used to build sieves of ever finer mesh. In this picture of science as progress through conjecture and refutations, there is no room for confirmation or corroboration.

VI

So much for pure science. What I wish to show next is that all this does not apply to applied science or to technological invention. In applied science there are two standard kinds of problems, deducibility and applicability. Given a theory and given a problem, the applied scientist asks himself, can I solve the problem while using that theory, and is my solution true? The first kind of question is essentially mathematical, and this explains why the terms 'applied science' and 'applied mathematics' are so often used interchangeably by people who will never use the words 'science' and 'mathematics' interchangeably. Mathematics, pure or applied, has its own kind of adventure of which I shall not speak beyond saying that it is the only kind of adventure an applied mathematician or an applied scientist may encounter. One may be even more specific here and add that in applied science, unlike pure science, the problem of deducibility is to find initial conditions which, together with given theories, yield conditions specified by practical considerations. This is, indeed, how Popper characterizes technology at large,[27] which is a rather narrow view of technology.

The second question – namely, is the solution to a given problem which he has deduced true – hardly allows itself of intellectual adventure, though designing the test may be, and executing it may also involve physical adventure. That the testing of results of applied science does not belong to applied science is rather obvious: even the applied scientists in industry, not to say in the universities, wear white collars, not white lab coats or blue overalls. They cannot and need not perform tests. That they cannot is a point of great significance, which I wish to label as 'Hatfield's law'. It is this: there is always a gap between applied science and the implementation of its conclusions, to be filled by invention.[28] The law is trivial in the sense that applied science does not issue programs for computers which implement its results; it is less trivial if we understand that applied science does not issue programs even for skilled workers without gaps to be filled by clever inventors.

To take a very simple instance[29] of the difference between invention and applied science, let us take a case involving both. The inventor Edison wished to replace gas street lighting with electrical lighting. The crux of the difficulty which disappointed other inventors was practical:

the amount of copper needed to conduct electricity along the streets of a city would be too large to warrant investment. If then occurred to Edison that this obstacle could be overcome if high tension and high resistance were employed. This was a technological idea, which could be correct or incorrect: Edison needed an applied physicist to tell him that, and his applied physicist, Upton, applied Ohm's law to the problem and concluded that by using a tension of 100 volts the quantity of copper needed could be cut into one one-hundredth of the originally calculated quantity required. The conclusion surprised even Edison, though it followed logically from a law he knew very well. Upton's job was finished on paper. Applied science is, thus, providing answers to given questions when these are implicit in a given theory. Posing the question and seeing the possible technological significance of the answer to it was one of Edison's inventions on the road to electric lighting.

Edison's opponents did not think the obstacle – the need for immense quantities of copper – could be overcome except by connecting the lights in series. Applying Ohm's law to this solution, they found it impracticable and showed that the light given by the many electric lamps needed to light a city would be too small to be of any use. Doubtlessly they were quite right in their applied science but rather wanting in imaginative invention. This shows how different are applied science and invention. That the two can overlap was shown by my example of the collaboration of Edison with Upton; that they are distinct is symbolized by the distinctness of these two individuals, their training, and dispositions.

One need not blame Edison's opponents, or any other people, for lack of imagination; for by doing so one may underrate the power of the imaginative inventor. Not only was Edison's idea so very new, it also required many other innovations to adjust to his new scheme. For instance, whereas dynamos in his days gave out constant currents, his idea required the invention of a usable dynamo with constant tension. It is conceivable that his idea of high-tension currents would not work if he could not invent such a dynamo. How much more understandable, then, it is that his opponents were thinking with existing dynamocs in mind. And this is not all. When all means for high-current lighting were found, there was no filament which could stand the conditions imposed by Edison well enough to be practicable. For all he knew, there may not have been such a filament. He himself conceded that much at the stage when he worked on platinum filaments which could not be produced commercially prior to the commercial development of machines to transmute base metals into platinum. It is incredible how lucky Edison was and how much he dared gamble on his good fortune.

Invention depends on finding facts for which we have no clues, at least a sufficiently wide absence of clues to make the haystacks in which they lie practically infinite. Otherwise, a team of applied scientists would find it, to be sure. And so the failure of a technical inventor is as final as the failure of a failed wildcat prospector.

Not seeing the wildcat character of invention, not seeing it as the miracle of finding a needle in an infinite haystack, often leads to misunderstanding and to underestimation of inventors and their sense of adventure. The most definitive and thorough history of photography is probably that of the Gernsheims.[30] Yet their view of Daguerre's invention is incredibly naïve; they claim[31] that Herschel was much greater than Daguerre, since after merely having heard about Daguerre's success Herschel repeated and improved in a few days the results that initially took years to develop. The truth is that there was no reason at all for anyone to think of the possibility of latent images to be developed. The problem of photography for decades had been how to fix an image which is visible at once (perhaps while turning it from negative to positive). No one thought of latent images to be developed, and everyone knew of visible images which soon disappeared. Thomas Young applied this knowledge of temporary images to photographing infrared interference – the first piece of photography of any value, and a typical case of applied science which is in no way an invention, as was Daguerre's thought of the theory of latent images to be first developed and then fixed. Daguerre had the thought under most peculiar circumstances, and the thought appeared surrealistic to him at the time and years after.[32] Once his idea proved successful, hosts of improvements were possible within a relatively short period.

It is because invention is a theoretical rather than a practical activity – though to a practical end, of course – that inventors may illustrate their inventions while using most crude machines; but it is because their claims for inventions must be corroborated that they must use some machines, unlike theoretical scientists and even applied scientists who may never bother about practice.

We see, then, that though corroboration is needed in technological invention, it is of no use in science, pure or applied. The question is whether corroboration has anything to do with belief, and I think I have shown that in technological invention corroboration has something to do with the nature of the problems at hand, which is always, how one can do successfully something specified; whereas in science, pure or applied, belief is not a matter of scientific method either. As Popper has stressed, in pure science we may try to refute the most corroborated view, and in applied science truth is of little importance. In applied science, that is, the question of the truth of the theories to be applied hardly even enters, though the question of the applicability of results from it is crucial and is answered by simple tests. Thus, we do not believe, but we still apply, Newtonian mechanics, though not to systems involving high velocities.

VII

Technology involves various factors, and I have thus far concentrated on only two of them, namely, applied science and invention. The implementation of the results both of an invention and its maintenance are important fields which I cannot discuss here beyond saying that in the field of implementation corroboration is most important, and the degree of corroboration required prior to implementation is socially and legally determined. It is doubtless that here the requirement for corroboration is a matter of public safety and that though no amount of corroboration insures success, the legal requirement for corroboration does eliminate some dangerous technical innovations prior to their implementation in the market. But if we demand much corroboration, there will be no implementation.

To show how complex the problem of implementation is, and how difficult, I should mention the classic trial of General Billy Mitchell, who thought he had corroborated his theory of the importance of air force bombers and the consequent diminishing importance of naval warfare. His sincerity is beyond question and is the main reason for the fame of his court-martial. Another reason for his fame is rather vulgar, namely, the truth of his prediction of an unannounced Japanese air attack. But, of course, he was mistaken in most of his views except about the power of air raids, and implementing all his proposals might have been a disaster.[33] Success philosophy is what makes the various books (as well as the movie) about him so very unsatisfactory. Without success philosophy the story of Billy Mitchell becomes much more interesting and enlightening, especially in relation to the requirement of the complacent – such as his superiors – for more corroboration before implementing any novelty. But I shall not go further into it. I wish only to say that it exemplifies one point of significance from our present viewpoint.

Those who believe that empirical evidence renders a theory credible view the implementation of that theory not as an adventure but, on the contrary, as the reduction of risk and thus success. In truth, the implementation of a novelty is risky even when no risks seem possible. Testing is performed so as to eliminate some risks – some 'bugs', to use Edison's jargon – and some risky innovations. But sometimes the prevention of one risky implementation entails the taking of other risks, as Mitchell's story shows. This argument has been used in the recent Senate subcommittee investigation into the disaster of the atomic submarine 'Thresher', which sailed in spite of known and eliminable risks. Whether the argument that too much testing and insisting on safety may lead to other risks was used properly or in order to shield comrades in arms I do not know, but such questions can be discussed rationally, and will be, perhaps, by future historians of technology.

As I have said before, success is not something to be proud of, but a puzzle to be explained. The discoveries of modern biology and modern physics about the hostile world in which we live should make our very survival a great puzzle. But, on the contrary, philosophers of science refuse even to see the obvious fact that luck and intuition are essential to success. Yet I must say this in their favor: When any phenomenon takes place regularly, we wish to explain it and thus systematize it. Discovery and invention look regular enough nowadays, and so they seem to require a systematization. That there is a flaw here I have tried to argue, but I cannot put my finger on it. The dichotomy between algorisms versus lucky intuition seems a fundamental fault in our way of thinking. Obviously, mixing algorism with intuition and luck will not destroy the dichotomy. One way out is to view systematization dynamically: some past successes which looked lucky can be explained by later systematization. This eases the pressure of the difficulty, but no more. Take any unsystematic but regular process: to use Chomsky's work,[34] take a child's intuitive learning to speak. Not only do we not know how to explain it, it seems as fundamentally mystifying as Faraday's or Edison's uncanny ability to discover and invent to order.

To conclude, the confusions in the field are rooted in a difficulty which is shared by all of us. Our dichotomy between algorisms and lucky inspirations is very unsatisfactory. Philosophy managed to progress in various directions in spite of this fundamental obstacle. But this obstacle may be a most serious impediment for the progress of the philosophy of technology. So, the field may be at a dead end for the time being, or, if it will get going, may lead to spectacular results which are bound to revolutionize all philosophy and much of our way of thinking in many departments.

Boston, Mass.

NOTES

* First published in *Technology and Culture* 7 (1966), 348–366.

[1] See my 'The Role of Corroboration in Popper's Methodology', *Australasian Journal of Philosophy*, XXXIX (1961), 82–91. See also my 'Science in Flux: Footnotes to Popper', in the third volume of *Boston Studies in the Philosophy of Science*. See also n. 22 below.

[2] K. R. Popper, 'Three Views Concerning Human Knowledge', *Contemporary British Philosophy* (ed. by H. D. Lewis), London, 1956, pp. 355–88; and *Conjectures and Refutations*, London and New York, 1962, pp. 97–119. See also my 'Duhem versus Galileo', *British Journal for the Philosophy of Science* VIII (1957), 237–48.

[3] Matthew Josephson, *Edison* (paperbound ed.), New York, 1959, pp. ix–xi.

[4] See my *Towards an Historiography of Science*, Beiheft 2, *History and Theory*, The Hague, 1963, pp. 8, 17, 42–48, 84, 85, 86, 105, 106.

[5] Pierre Simon, Marquis de Laplace, 'Application of the Calculus of Probabilities to Moral Philosophy', *A Philosophical Essay on Probabilities* (paperbound ed.), New York, 1951, Chap. X.

[6] Josephson, *Edison*, p. 198.

[7] *Ibid.*, p. 216.

[8] Rudolf Carnap, *Logical Foundations of Probability* (2d ed.), Chicago, 1962, p. 193.

[9] 'A General Estimation Function', in *ibid.*, Sec. 99 ff. See also his 'On Inductive Logic', *Philosophy of Science* XII (1945), Sec. 10 (reprinted in *Probability Confirmation, and Simplicity* (ed. by M. H. Foster and M. L. Martin), New York, 1966). Cf. Popper, 'On Carnap's Version of Laplace's Rule of Succession', *Mind* LXXI (1962) 69–73; and my 'Analogies as Generalizations', *Philosophy of Science* XXXI (1964), 351–56.

[10] J. Wolfowitz, 'Abraham Wald', *Annals of Mathematical Statistics* XXXIII (1952), 9: "... was put on the restricted category and made available only to authorized recipients. Wald chafed greatly under this restriction." Cf. Statistical Research Group, Columbia

University, *Selected Techniques of Statistical Analysis for Social and Industrial Research and Production Management*, New York, 1947, pp. viii–ix.

[11] William Whewell, *The Philosophy of the Inductive Sciences*, London, 1840, Vol. II, Part 2, Book XI, Chap. i, Secs. 6, 10; Chap. v, Secs. 7, 8 (about testing), 10 (about techniques of testing: predictions), 11 (about the stringency of the test: new prediction), 12, 13 (correlating simplicity and testability, so to speak); Chap. vi, Sec. 12 (non-*ad-hoc*ness). See also his *History of the Inductive Sciences* (3d ed.), London, 1843, Vol. I, Book VII, 'The Discovery of Neptune': "Thus to predict unknown facts found afterwards to be true, is, as I have said, a confirmation of a theory which in impressiveness and value goes beyond any known explanation of known facts." And such confirmation, he says, took place only a few times in the whole history of man.

[12] Popper, *The Logic of Scientific Discovery*, New York, 1959, Chap. x and New Appendix, IX.

[13] Josephson, *Edison*, p. 198.

[14] See my 'Science in Flux' (n. 1 above), for a detailed discussion of this statement.

[15] Michael Polanyi, *Pure and Applied Science and Their Appropriate Forms of Organization*, Society for Freedom in Science, Occasional Pamphlet No. 14, Oxford, December 1953, p. 2. See also p. 9.

[16] For Polanyi, *see ibid.*, p. 6, on the ability of the competitive system to cope with technological novelties. See also his *Personal Knowledge* (paperbound ed.), New York, 1964, p. 177 n. See n. 17 for the exactly opposite view. For the Selden case, see William Greenleaf, *Monopoly on Wheels: Henry Ford and the Selden Automobile Patent*, Detroit, 1961, and the review of that volume by John B. Rae in *Technology and Culture*, II (1961), 289. For Edison's 'My 40 years of Litigations', *Literary Digest*, Sept. 13, 1913, see Josephson, *Edison*, pp. 354–60, and his Index article 'Patent Infringement'.

[17] For the Baconian tradition, see my *Historiography* (n. 4), pp. 4–6, 15, 60–66, 81. For the Malinowski tradition, see Ian C. Jarvie, *The Revolution in Anthropology*, London and New York, 1964, pp. 1–7, and his Index article 'Fieldwork'.

A. P. Usher, *A History of Mechanical Invention* (rev. ed.), Cambridge, Mass., 1954; paperbound ed., Boston, 1959, p. 68: "Cultural achievement is a social accomplishment based upon the accumulation of many small acts of insight of individuals. The massiveness of this social process was long ignored or misunderstood. ... A conspicuous result of this disposition to put a part for the whole was the frequency of bitter controversies over claims of various inventors to a particular invention .... These disputes all rest on the false assumption that the achievement was so simple and specific that it could properly be identified with the work of a single person at a single moment." A clearer and sharper statement is quoted from seminar notes by Thomas A. Smith in his enthusiastic review of Usher's volume, in *Technology and Culture*, II (1961), 36. "The concept of heroes and men of genius," says Usher according to Smith's record, "are literary and cult devices that simplify the historical record." See also paperbound edition, p. vii. For my view of the partial justice of such wild claims, see my *Historiography* (n. 4), pp. 31–33.

[18] Pierre Duhem, *The Aim and Structure of Physical Theory* (paperbound ed.), Princeton, N. J., 1962, Part 1, Chap. ii; Part 2, Chap. iii; and Appendix 'The Value of Physical Theory'.

[19] E. Mach, 'The Economical Nature of Physical Research', *Popular Scientific Lectures*, La Salle, Ill., 1907, Chap. xiii; his *The Science of Mechanics* (paperbound ed.), La Salle, Ill., 1960, Chap. iv, Sec. 4.

[20] Popper, 'Three Views' (see n. 2), Sec. 5: 'Criticism of the Instrumentalist View'.

[21] Erik H. Erikson, 'Autonomy vs. Shame and Doubt', *Childhood and Society* (2d ed.), New York, 1963, Chap. vii, Sec. 2, p. 253; Allen Wheelis, *The Quest for Identity*, New York, 1958, Chap. vi, esp. pp. 183, 199, 201, 250.
[22] Popper, *Conjectures*, Chap. x, Sec. 23, and note on p. 248.
[23] J. M. Keynes, *Treatise on Probability*, Cambridge, 1921, Chap. xxiii, esp. p. 271.
[24] Popper, *Conjectures*, p. 217: "I assert that we *know* what a good scientific theory should be like, and – even before it has been tested – what kind of theory would be better still, provided it passes certain crucial tests." And on p. 242, "The second requirement ensures that...."
[25] Popper, 'The Aim of Science', *Ratio* **I** (1957), 24–35. See also my 'Between Micro and Macro', *British Journal for the Philosophy of Science* **XIV** (1963).
[26] See my 'Sensationalism', *Mind* **LXXIV** (1966), 1–24, for a detailed discussion of this statement.
[27] Popper, 'Naturgesetze und Wirklichkeit', in *Gesetz und Wirklichkeit* (ed. by S. Moser), Innsbruck, 1949, pp. 43–60, esp. pp. 53 ff.
[28] H. Stafford Hatfield, *The Inventor and His World*, 1933; Pelican ed., 1948, pp. 111, 133, 134, 151, *et passim.*
[29] Josephson, *Edison*, Chap. x, esp. p. 194. Cf. Usher, *A History of Mechanical Invention*, pp. 72–77 and 401–6.
[30] Helmut Gernsheim, in collaboration with Alison Gernsheim, *The History of Photography from the Earliest Use of the Camera Obscura in the Eleventh Century up to 1914*, Oxford, 1955.
[31] *Ibid.*, p. 81.
[32] Helmut and Alison Gernsheim, *L. J. M. Daguerre: A History of the Diorama and the Daguerreotype*, London, 1955; American ed., published under title *L. M. J. Daguerre, the World's First Photographer*, Cleveland, 1956.
[33] Ruth Mitchell, *My Brother Bill: The Life of General 'Billy' Mitchell*, with an Introduction by Gerald W. Johnson, New York, 1953, pp. 12–13: "There was a moment when he seemed to be entirely right. But events in Korea, where our separate airforce has accomplished nothing ... prove .... But even if he had been wholly wrong ...."
[34] Noam Chomsky, 'Explanatory Models in Linguistics', in *Logic, Methodology, and Philosophy of Science* (ed. by E. Nagel, P. Suppes and A. Tarski), Proceedings of the 1960 International Congress, Stanford, Calif., 1962, pp. 528–50.

J. O. WISDOM

THE NEED FOR CORROBORATION

*Comments on J. Agassi's Paper**

Joseph Agassi's thesis is that applied science must be distinguished from technology. And he brings two considerations to bear on this.

(1) Agassi lumps together induction, confirmation, and corroboration as being *algorithmic*. An interpretation of science that is algorithmic, indicating that scientific achievements can be or are reached mechanically, runs counter (a) to the fact that science is an adventure and (b) to the inventiveness or imagination required in doing science. Thus, according to Agassi, corroboration does not characterize science – that is, theoretical or pure science. Nor does he consider that it characterizes applied science. But, he holds, it does characterize technology. Hence applied science is different from technology, and the difference lies in corroborability.

(2) Agassi despises success in a big way, not only in the form of status symbols, but even the successful predictions of pure science. Such prediction, or fortune-telling, is corroboration; corroboration is not needed by science; therefore, successful prediction is not. Nonetheless technological success is permitted; or, rather, success is appropriate to technology. Hence, once more, applied science differs from technology, and the difference concerns the relevance of success.

Thus on both counts, applied science (as well as pure) differs from technology; and the cornerstone of the distinction is that corroboration is not appropriate to science, pure or applied, but is appropriate to technology.

To this Agassi adds the interesting point that there is a gap between applied science and its implementation, and that this gap is filled by invention. To all this, Agassi adds a problem he cannot solve.

With a great number of points I agree, and I shall not mention them; I shall take up only a few points that lack cogency, and expatiate a little on the main thesis.

(1) First, about success: Agassi's view that scientific predictive success is valueless seems to be based on a misconstruction of Popper. Agassi

rightly points out that the greatness of a scientific theory, in Popper's view, transcends success; for Popper brought out the importance of false theories – the fact that they could be false elevated them above what could not be falsified. Now Agassi seems to interpret this as meaning that only falsifiability matters, that it, that there is no gain when a theory survives an attempt to falsify it. Agassi would seem to be saying that all that counts is refutability (or predictability to obtain refutability) and not correct prediction, that is, that it is a matter of indifference whether a prediction is true or false – and perhaps there is the innuendo that falsification is more to be valued.

With this I do not agree, and I do not think it is Popper's theory. In Aggasi's account, Einstein's theory of general relativity would have been just as valuable even if Eddington's eclipse observations had gone against it. I agree that even if they had, the theory would have been a great one; but I consider it was all the greater because they failed to falsify it. It is easy to express Popper's view in the provocative form that the aim of a test is to refute a theory. But one must add that, valuable as a refutation would be, a failure to refute would be more valuable, and that when a test is designed, not to refute, but to be able to refute a theory, one hopes the refutation will fail.

Several reasons may be given for this. (a) The successful prediction, which could have falsified, gives us a better explanation of something than we had before and new knowledge about the world. (b) We want to extract the most out of a theory before we lose it by eventual refutation. (c) When we have a problem and construct a theory to solve it, what we want to do is just to solve the problem – not to *fail* to solve it!

(2) Agassi takes Popper to task for saying that with a new theory it is beneficial to have some corroboration early in its life and not to have it refuted at once. Lest this possibly *sound* like verification or pandering to a scientist's need for a little bolstering up, nonetheless it contains a serious point which should be given adequate methodological expansion (without prejudice to what Popper may have had in mind). Agassi says that falsification of a theory should not lead us to withdraw our appreciation of it. Now if it is falsified right away, we shall not explore it further, and there may be several highly important corroborations we shall never know about. Hence our appreciation will be too low, not fust because of being deprived of successful predictions, but because of

failure to know of its wide explanatory power; and this will not be due to our weakness in valuing success, but because we are cut off from the possibility of exploring the theory (unless an extraordinary new research policy is introduced of pursuing investigations into refuted theories). Thus we need to get the greatest volume of corroboration out of a theory it is capable of.

Looking at the matter from the opposite angle, Agassi holds that a theory, even if refuted at once, is valuable. But if a theory has received *no* new corroboration, that is, provides as much explanation as an already existing theory, I do not see that it is to be valued. (It is an exercise in ingenuity, but is it more?) Hence, to be valuable it needs corroboration not refutation at the first test.

(3) Agassi's unsolved problem is this. The great success of science suggests we might reasonably seek a way of understanding how it is done, and this would reduce scientific discovery to an algorithm, thus depriving science of its adventurous nature and dispensing with the need for inventiveness and imagination. I am not impressed with this doubt.

He and I would agree that to teach Popper's methodology to students of the social sciences would almost certainly lead to improved results (theories, hypotheses, explanations) and a greater number of them. This would not diminish the imaginative qualities at all; it would increase them by harnessing them into a usable form. Supposing Schon's idea[1] – that discovery is the displacement of an idea from one context to another – the realization of this might well lead to better results, for researchers would look for ideas to displace and for contents to displace them to. But this (algorithm?) would not tell them *what* ideas to use or *where* to focus them, and the old imaginative qualities would be needed undiminished.

(4) Agassi said little about what discriminates applied science from technology. I agree they are different, and I wish merely to reinforce this and add a small point.

The discovery of H_2O structure was science; that of 606 or M&B 693 technology. The difference is not *in rerum natura* but in aim: the one to understand structure, the other to create a structure for a certain purpose.

I may mention some of Kuhn's excellent examples[2] with others of the same ilk. The application of Newtonian mechanics to resisting media is applied science; if the medium is highly specific, so that we take a special interest in it (such as water because we want to fire torpedoes in it), we move into technology. The one is concerned with understanding and extending knowledge, the other with using it. Calculating the mass of the moon or densities is applied science, and the information may be needed in theoretical science or in technology, either for testing Newton's theory, for example, or for landing on the moon smoothly. If you hang up a heavy flexible string, you apply Newtonian mechanics so as to find the equation of the catenary. This again is extending knowledge or understanding.

It seems to me that an already known distinction holds, though firmly in need of refocusing. Science is to understand, technology to do. But applied science, though a step on the way to do something, is itself an extension of understanding. Applied science has sometimes been described as concerned with doing, which seems to me to be wrong in linking it with technology rather than with (pure) science.

Though I have disagreed with some of Agassi's means of getting there, I agree fundamentally with his conclusion that applied science is distinct from technology.

Los Angeles, Calif.

NOTES

* First published in *Technology and Culture* 7 (1966), 367–370.

[1] D. A. Schon, *Displacement of Concepts*, London, 1963, Part 1.

[2] T. S. Kuhn, *The Structure of Scientific Revolutions*, Chicago, 1962, Chap. iii.

J. AGASSI

PLANNING FOR SUCCESS

*A Reply to J. O. Wisdom**

Professor Wisdom agrees with me[1] that applied and science technology are distinct, but for different reasons. Since 'reason' may be either 'criterion' or 'end', let me make it clear that I use a simple factual criterion to distinguish the two, and to a definite philosophical end. Consider the following inference: Corroboration (or confirmation or verification, or factual support, or agreement with experience) is important in technology; technology is identical with science; *therefore*, corroboration is important in science. The inference is valid; its conclusion is false; and its first premise is true; hence, the identification of science and technology is mistaken. This suffices for me; for those who do not agree that the conclusion is false, and even insist that it must be true because the two premises are true, for them different arguments might be of use. If they see that the identification of science and technology is erroneous independently of the theory of corroboration, then they might be more agreeable to reform their ideas of corroboration too.

It is a simple matter of fact, I think, that applied scientists use high-powered theories, difficult mathematics, etc., to solve problems in a way not open to those less well versed in these theories and deductive techniques; technologists are not usually as theoretically high powered and do not work out patiently and carefully chains of deduction – they hire applied scientists for that – but they are high powered in different respects, in being more familiar with technological situations, facts, problems, tasks, etc., from the availability and specific merits and defects of raw materials, to storage, to marketability. My example of a technologist and his work was Edison and his building of an electric street-lighting system; this involved, *inter alia*, invention, applied science, organization of implementation, economics, and even corporation law. When it came to applied science, Edison used an applied scientist – Upton – to make the calculations. Whether the results of the calculation were correct was not Upton's concern but Edison's – as long as the deductions were valid, of course. Applied scientists do not usually wear lab coats (except perhaps for the sake of prestige) as do experimental scientists and technicians – including engineers and inventors. When the New York electric system was inaugurated, Edison was there, tailcoat and all. Soon troubles ensued, and he quickly tried to straighten them out. In no time, he looked as if he were wearing the previous year's overalls. Excitement of this kind may take place in technology and in experimental science; excitement in theoretical science and in applied science is very different.

Since Wisdom kindly clarifies my views, I wish to take leave to clarify his clarification on two points: success and scientific success. There is no objection in my paper to success, but only to success philosophy, to the view of success as the measure of worth. The better the society and the more favorable the circumstances, the more likely, perhaps, that success and worth would go hand in hand; that is the most one could reasonably say. Anyway, our society is not that just; not as yet, at least.

As to scientific success, I dislike the idea that it can be generated by a formula or an algorism, and I dislike the idea that it can be assessed by a formula. I have given an example of a formula pertaining to performing both tasks – Laplace's – and I have given an example for a formula pertaining to performing only the second task but *not* the first task – the Whewell-Popper formula as generally understood. I am not clear whether this point of mine is correctly presented by Wisdom.

The formula for corroboration which is not an algorism is the Whewell-Popper formula: A theory is corroborated if, and only if, after it has been invented and shown to explain the puzzling phenomena it was put to serve, it has been tested in an attempt to refute it, and that attempt has thus far failed to refute it.

I gave an example of how the Whewell-Popper formula would be misused as a measure of Einstein's worth because it would make Einstein's worth depend on luck concerning matters entirely beyond his control, such as the Eddington experiment. Wisdom shows quite clearly – and in my view correctly – that this does not apply to other factors, which were very much within Einstein's control, such as knowledge of experimental results attained *prior* to his studies. True, Einstein was encouraged by his ability to explain the motion of the perihelion of Mercury;

this is irrelevant to my criticism of the use of the Whewell-Popper formula as a measure of worth, as it merely purports to show that in some cases such a use of the formula is not as objectionable. Also Wisdom is mistaken here on a technicality. Both Whewell and Popper would not consider the Mercury perihelion motion a corroboration, but they would consider Eddington's experiment a corroboration.

Wisdom fears, incidentally, that if a theory is refuted at the first test it will not be tested further, and so experiments in agreement with it might not be discovered. Perhaps. Also, let me add, if a theory is not refuted at first test, then possibly it will not be tested further and refuted – sometimes even with premature implementation, leading to refutations descending on us as catastrophes. There are four logical possibilities here (early corroborations or refutations giving or not giving false impressions), and there exist historical examples conforming to each of them. So arguments here go neither way, except, perhaps, in some limited historical contexts. But, as we are concerned here neither with history nor with psychology, arguments go neither way. My view that qua corroboration, corroboration cannot play any role in science was drawn from Popper's philosophy. A critic may show that my reasoning was faulty or that Popper's philosophy is false; if he denies either possibility he is no critic of my view.

But Wisdom is right in claiming that we do wish to know both the facts which agree with a given theory and those which do not. This is because we wish to know how much a theory explains; and the more it explains some puzzles, the better. Now suppose that we go back two or three centuries and find there a theory which explains a phenomenon discovered one or two centuries ago by one who had no knowledge of that theory; and suppose, further, that the phenomenon is not problematic today. Then, the present discovery that one old fact agrees with a somewhat older theory is of no interest. Why? Popper would say, because no one intended to refute that theory by that experiment, and so it is no corroboration. Scholars have been puzzled by Popper's stress on intention, and not without reason. I think intentions as such need not matter, certainly not the sincerity of the attempt to refute, which regrettably Popper stresses so much. But intentions may enter when they are more objective, such as the desire to solve certain problems. When we wish to know how much a theory explains, and wish it to explain much, surely a corroborating experiment does that. When we wish to find, if a theory is false, that it is false, then corroborations will not do but refutations will. No one wants his partner to cheat him; but some people want to know if their partners cheat them, when, and how they cheat them. Some do not want such information even if it is true, and I tend to feel like them. But I do want to know where my theories went wrong if they are not the absolute truth. It all depends on what we are after.

And if we are after solutions to certain problems and we have a sure method of solving them, we shall, of course, indorse this method; even if the method is only partially successful (in any sense) we shall indorse it. To speak against success here is folly. But suppose all our problems were easily solved and life were deprived of all challenge. Paradoxically, success would become failure by excess. The literature on the poor rich is maudlin and vulgar, but there is a kernel of truth in it.

When we learn to operate a machine or an algorism we may feel challenged, but when we master it we cannot enjoy a life consisting chiefly of operating it, no matter how successful the operation is in terms of economic and social standings or in any other terms. This is a psychological truth concerning youngsters and adults, humans and even some other animals.

If we feel challenged when we learn to operate a machine or an algorism, but not when operating it, then such a progress is not mechanical or algoristic. If so, how is it that practically everyone learns to operate the algorism in the same way? We all learn our mother tongue to an incredible uniformity. Einstein's theory, being non-formal, may be open to myriads of understandings or misunderstandings. Yet most men of science understand it in one, two, or three standard interpretations. I think this is an interesting fact. It may have to do with the fact that Einstein's theory came primarily to answer some questions, and that somehow questions have a logic of their own which is capable of directing people again and again toward understanding neither by mere mechanics nor by mere luck and ingenuity which defy generalization. My difficulty is exactly this. Suppose that statements about intuition can be generalized like statements about other operations. Then intuition may feel different from memorizing by rote, but it is not: Then, we can have a sort of intellectual progress by rote, a sort of science-making machine. If intuition is so unique, if we have no science-making machine, how is it that science progresses so steadily? This question seems to be of some significance in the philosophy of technology: As Dr. Jarvie has pointed out, the American space program is based on an estimated rate of technological progress. This optimism strikes me as most incredible. Can we explain it?

Boston, Mass.

NOTES

* First published in *Technology and Culture* **8** (1967), 78–81.

[1] Professor Jarvie's paper referred to in the end of this paper is printed in *Technology and Culture* 7 (1966), 371–383 [see this volume]. Since all references in this paper are to my previous one, I find no need for further references, except where I speak of objectivity of problems. To my knowledge there is no discussion of this point anywhere except in Sir Karl Popper's recent *Clouds and Clocks*, St. Louis, 1966, Section XX.

J. O. WISDOM

RULES FOR MAKING DISCOVERIES

*Reply to J. Agassi**

Joseph Agassi's paper and my critical appraisal were not very far apart; his reply shows we are closer than appeared. He notes we both accept the distinction between applied science and technology, though for different reasons; he agrees about the need to find out what agrees and what does not agree with a theory and that we want to solve our problems; he corrects me about success, for the attacks not success but success philosophy. (I confess I still think that in denying the relevance of corroboration to science he is denying the significance of success.) Our slight divergences are these: (1) Agassi argues against my estimate that there might be grave loss if a theory were refuted at the first test. (2) He thinks I have made a technical mistake about what is to count as corroboration on Popper's theory. (3) Agassi continues to object to reducing scientific invention to an algorithm.

(1) Agassi does not deny my point that there might be a loss if a theory were refuted early but counters by maintaining that possible loss might arise if a theory that survived its first test were not tested again. In terms of the attitude that would arise among scientists, it would be natural to drop a theory that had had one refutation but also natural to continue testing a theory that had survived, and in addition the work of exploiting and applying a tested theory could lead to further testing and refutation. So here Agassi seems to be wrongly assessing ordinary realities.

(2) I counted the motion of Mercury's perihelion as a corroboration of general relativity, and Agassi says that on Popper's methodology this is a technical error, for a corroboration comes after the construction of the theory. But an event providing corroboration could certainly come before this, at least if it was not what the theory was devised to do. Einstein worked in a very unusual way in his relativity thinking, and I have the impression that he was not trying to think up a theory directly that would explain Mercury but was trying to deal with problems about mass and emission of light and gravitation and that he was imbued strongly

with a methodological prescription involving two connected requirements, one being that the form of a law should remain invariant with regard to frame of reference and the other having to do with the metric; for his belief seems to have been that if you could get a new theory in this way it would be sure to do some explaining. I would call his three theories, not relativity, but theories of invariant metrics. So it was just a beautiful bonus when general relativity was found to explain Mercury. And Mercury therefore did constitute a corroboration.

But even if this account does not accord with Einstein's aim, I fail to see why an event that a theory is constructed to explain should not provide corroboration when the theory is found to explain it; for corroboration is failed refutation, and the event could have refuted the theory.

(3) As regards reducing invention to an algorithm, Agassi is in effect maintaining that if you can make a rule for making discoveries, you *ipso facto* nullify the possibility of making a discovery, for your rule conflicts with the spontaneous creativity required.

Perhaps Agassi does less than justice to the degree of specificity involved. In *Waiting for Godot*, Lucky is ordered to *think*. Suppose he is told to think logically or validly. This restricts how he shall think but leaves a lot of scope. If he is also told to think testable hypotheses, this is a further restriction but leaves endless scope for imagination. I have grave doubts whether an absolutely specific rule for obtaining an absolutely specific discovery could be obtained, and perhaps that is all Agassi wants. But it *would* at least be possible, in my view, to have rules for breaking down certain sorts of tasks to such small dimensions that it is a moral certainty that some able technologist will find some way or other of accomplishing them. We may be able to reduce the task of discovering a new drug to have certain required properties to such simple dimensions that mediocre students, though not a robot, could do it.

But supposing an algorithm of discovery is possible, what then? We could always use our imagination instead when we wanted to exercise it. For one of the most beautiful methods of obtaining rate of change in mathematics, namely, differentiating, there exists an algorithm. So when we are in a hurry we are not obliged to enjoy the pleasure of differentiating; but we can do so at will. The fact that artificial insemination may well be more productive does not necessarily mean that conjugal relations will die out.

The interesting question, of course, is whether an algorithm of discovery is impossible in principle. I am sceptical of all philosophical a priori attempts (Humean, Kantian, positivist, linguistic) to show that something is impossible. Factual impossibility is not Agassi's concern. It is hardly simply a self-contradiction – that would be too simple. With what scientific theory does it conflict? If there is one, that theory may turn out to be false. The only basis left would seem to be that it is synthetic a priori, which would be an unholy alliance for Agassi to accept.

Los Angeles, Calif.

NOTE

* First published in *Technology and Culture* **8** (1967), 210–212.

H. SKOLIMOWSKI

THE STRUCTURE OF THINKING IN TECHNOLOGY*

Inquiry into the philosophy of technology, due to the infancy of the subject, must start with some reflections on what technology itself is. There is at present a tendency to identify technology with a demiurge or our times, or perhaps even with a Moloch who will bring doom to mankind, that is, mankind as dreamt of by philosophers, not by organization men. In this setting technology assumes a role similar to that which was ascribed to history in the nineteenth century: the role of the final cause which shapes the destiny of mankind and, more specifically, which aims at the total subjugation of man to the machine or, in other words, at turning the human being into a technological component.

It cannot be denied that reflections on technology in this fashion are philosophical reflections and that consequently they belong to some system of the philosophy of technology. At this point, however, a vital distinction should be made between a *philosophy of technology* and a *technological philosophy*. The former belongs to the realm of epistemological inquiry and attempts to situate technology within the scope of human knowledge; the latter belongs to the realm of sociology, broadly conceived, or social philosophy, and is concerned primarily with the future of human society.

Those who prophesy that our civilization will be devoured by the Moloch of technology are expanding a certain vision of the world, are viewing the world through technological lenses, are attempting to establish a new kind of monism, the technological monism, in which the technological order is shown to be the prime mover and the ultimate justification of other orders, moral, aesthetic, cognitive, social, and political. The articulation of this technological philosophy is perhaps most important from a social point of view – as a way of alerting us to the dangers of technological tyranny. However, for the time being this technological monism, or whatever name is given to this sociohistorical prophecy, is but a prophecy. As important as it may be from a human point of view, it cannot serve as a substitute for a philosophy of technology proper, that is, for a philosophy that aims at the investigation of the nature and structure of technology, conceived as a branch of human learning and analyzed for its cognitive content.

I shall not be concerned here with the transformation of society by technology. It seems to me that the 'monolithic technical world' is but a graphic and perhaps fearsome expression, but not reality. For the time being the evidence that technology pervades the totality of human relationships is rather slim. In the realm of art, for example, modern technology perpetuates at least some traditional human values. The unprecedented spread of superb reproductions of the great masters, the easy availability of the finest recordings of music of the last five centuries, the spectacular rise in the production and distribution of paperback books, are all due to the advances of technology, and all serve, at least in part, the cause of highbrow culture, not technological culture.

It may be that a *comprehensive* philosophy of technology should include the moral implications of technological progress. It may be, as some philosophers insist, that, in spite of the semiautonomous development of technology, a substantial part of modern technology is moved by non-technological forces, that, for example, motor cars are produced in order to make money, intercontinental missiles in order to kill people. Consequently, a comprehensive treatment of the philosophy of technology must examine the presuppositions lying at the foundation of these technological 'events' and must a tempt to assess their implications for mankind at large. The weight of these problems cannot be underestimated. However, they are outside the scope of my considerations.

In this paper I shall be concerned with what I call the philosophy of technology proper, that is, with the analysis of the epistemological status of technology. Technology is a form of human knowledge. Epistemology investigates the validity of all human knowledge, its conditions, its nature. Therefore, it is the business of epistemology to investigate the peculiarities of technology and its relation to other forms of human knowledge. In particular, it is of crucial importance to analyze the relationship of technology to science. I shall argue in the course of this paper that: (1) it is erroneous to consider technology as being an applied science, (2) that technology is not science, (3) that the difference between science and technology can be best grasped by examining the idea of scientific progress and the idea of technological progress.

In the following sections I shall attempt to provide a basis for a philosophy of technology rooted in the idea of technologial progress. Then I shall proceed to show that in various branches of technology there can be distinguished specific thought patterns which can be seen as explaining technological progress.

Many methodologists and philosophers of science insist that technology is in principle a composition of various crafts. Regardless of how sophisticated these crafts may have become, they are still crafts. It is argued that technology is methodologically derivative from other sciences, that it has no independent methodological status, and that what makes it scientific is the application of various other sciences, natural sciences in particular. Thus, the scientific part of technology can be decomposed into particular sciences and accounted for as physics, optics, chemistry, electromagnetics, etc. This view misconstrues the situation because it does not take into account the idea of *technological progress*.

My thesis is that technological progress is the key to the understanding of technology. Without the comprehension of technological progress, there is no comprehension of technology and there is no sound philosophy of technology. Attempts that aim at reducing technology to the applied sciences fail to perceive the specific problem situation inherent in technology. Although in many instances certain technological advancements can indeed be accounted for in terms of physics or chemistry, in other words, can be seen as based on pure science, it should not be overlooked that the problem was originally not cognitive but technical. With an eye to solving a technical problem, we undertake inquiries into what is called pure science. Our procedures are extremely selective. Out of infinitely many possible channels of research only very few are chosen. Problems thus are investigated *not* with an eye to increasing knowledge but with an eye to a solution of a technical problem. If it were not for the sake of solving some specific technological problems, many properties of physical bodies never would have been examined, and many theories incorporated afterward into the body of pure science never would have been formulated. Perhaps the most obvious examples can be found in the sciences of electronics and of space physics. The development of computors resulted in the replacement of tubes by transistors. In developing transistors many properties and laws governing the behavior of semiconductors have been formulated which might never have been formulated otherwise. To take another example, the problem of metal fatigue and many other phenomena concerning the behavior of solids in space might never have been investigated, and theories resulting from them might never have been established if it were not for the sake of constructing supersonic planes and intercontinental rockets. To mention finally atomic physics, it was in the Manhattan Project where plutonium, an element not found in nature, had to be developed in the process of producing the atom bomb. Thus, in one sense science, that is pure science, is but a servant to technology, a charwoman serving technological progress.

I shall now discuss the thesis that technology is not science. By this statement I mean to say that the basic methodological factors that account for the growth of technology are quite different from the factors that account for the growth of science. Consequently, the idea of technological progress as contrasted with scientific progress must be examined more carefully.

I am in full agreement with Karl Popper that science, in order to exist, must progress; the end of scientific progress is the end of science. This progress results from the continuous improvement of scientific theories and constant enlargement of the scientific store; more precisely it results from a permanent overhaul of theories and incessant replacement of worse theories by better ones; 'better' means simpler, or more universal, or more detailed, or of greater explanatory power, or all these things together. The objective underlying this endless succession of theories is the increase of knowledge. The pursuit of knowledge (which is another expression for the pursuit of truth) has been and still is the most important aim of science. We critically scrutinize our theories by devising tests of increasing ingenuity and severity in order to learn how squarely they can face reality. Whatever operationists and conventionalists of various denominations may say, science is about reality. The acquisition of knowledge and the pursuit of truth are only possible if there is reality. Thus it is contained in the idea of scientific progress that we investigate reality and that we devise theories of increasing depth in order to comprehend this reality.

What about technology? Is it another instrument for investigating reality? Does it aim at the enlargement of knowledge and the acquisition of truth? The answer is negative in both cases. Hence we come to significant differences between science and technology. In science we *investigate* the reality that is given; in technology we *create* a reality according to our designs. In order to avoid confusion I should perhaps say at once that these two kinds of reality are not of the same order. To put it simply, in science we are concerned with reality in its basic meaning; our investigations are recorded in treatises 'on what there is'. In technology we produce artifacts; we provide means for constructing objects according to our specifications. In short, science concerns itself with what *is*, technology with what *is to be*.

The growth of technology manifests itself precisely through its ability to produce more and more diversified objects[1] with more and more interesting features, in a more and more efficient way.

It is a peculiarity of technological progress that it provides the means (in addition to producing new objects) for producing 'better' objects of the same kind. By 'better' many different characteristics may be intended, for example: (a) more durable, or (b) more reliable, or (c) more sensitive (if the object's sensitivity is its essential characteristic), or (d) faster in performing its function (if its function has to do with speed), or (e) a combination of the above. In addition to the just-mentioned five criteria, technological progress is achieved through shortening the time required for the production of the given object or through reducing the cost of production. Consequently, two further criteria are reduced expense or reduced time, or both, in producing an object of a given kind.

It hardly could be denied that contemporary freeways and highways mark a technological advancement in terms of durability when compared with Roman or even nineteenth-century roads; that modern bridges are far more reliable (in addition to other advantages) than bridges of previous centuries; that photographic cameras installed in artificial satellites are considerably more sensitive (in addition to being more reliable and more durable) than those used in the pre-Sputnik age; that the speed of jet airplanes makes them superior to the planes of the brothers Wright. And no one can deny that if the same plane or bridge or camera can be manufactured less expensively, or alternatively in shorter time (at the same expense), then it will equally mean a technological advancement.

The criteria of technological progress cannot be replaced by or even meaningfully translated into the criteria of scientific progress. And, conversely, the criteria of scientific progress cannot be expressed in terms of the criteria of technological progress. If an enormous technological improvement is made and at the same time no increase in pure science is accomplished, it will nevertheless mark a step in technological progress. On the other hand, it is of no consequence to pure science whether a given discovery is utilized or not; what is of significance is how much the discovery adds to our knowledge, how much it contributes to the comprehension of the world.

It may be argued that in the pursuit of technological progress we often bring about scientific progress as well. It should be observed, on the other hand, that scientific progress may and indeed does facilitate technological progress. Discoveries in pure sicence, regardless of how abstract they appear at first, sooner or later find their technological embodiment. These two observations lead to a conclusion that perhaps neither scientific nor technological progress can be achieved in its pure form; that in advancing technology, we advance science; and in advancing science, we advance technology. This being the case, it should not prevent us from analyzing these two kinds of progress separately, particularly because scientific progress is often treated autonomously and is regarded as the key to an explanation of the growth and nature of science. If we are permitted to divorce scientific progress from technological progress when examining the nature of science, we should be equally permitted to divorce technological progress from scientific progress when examining the nature of technology.

In this context it is rather striking that even such mature and eminent philosophers of science like Popper have nothing better to say than to equate technology with computation rules. Neither Popper nor, to my knowledge, any other authority in the philosophy of science, has cared to examine the idea of technological progress. Hence their remarks on technology, whenever they find it convenient to mention it, are rather harsh and far from adequate.

To summarize, scientific and technological progress are responsible for what science and technology, respectively, attempt to accomplish. Science aims at enlarging our knowledge through devising better and better theories; technology aims at creating new artifacts through devising means of increasing effectiveness. Thus the aims and the means are different in each case.

The kernel of scientific progress can be expressed simply as being the pursuit of knowledge. The answer seems to be less straightforward with regard to technological progress. However, in spite of the diversity of criteria accounting for the advancement of technology, there seems to be a unifying theme common to them all, or at any rate into which they can be translated. This theme is the measure of effectiveness. Technological progress thus could be described as the pursuit of effectiveness in producing objects of a given kind.

Now, the question is: Can this measure of effectiveness be studied in general terms or, to put it differently, can we aim at a general theory of efficient action and then incorporate it in the idea of technological progress? And a second question: Is there only one, or are there many different patterns leading to an increase of the measure of effectiveness in different branches of technology?

In relation to the first question, it should be observed that, in addition to specific formulas for efficient action constructed for limited scopes of human activity (e.g., the science of management), there is indeed a general theory of efficient action for all activities we choose to analyze. This general theory of efficient action is called *praxiology*. This theory has been worked out in detail by the Polish philosopher, Tadeusz Kotarbiński. Since the principles of praxiology are treated extensively in Kotarbiński's treatise,[2] I shall be very brief here.

Praxiology analyzes action from the point of view of efficiency. Praxiology is a normative discipline; it establishes values, practical values, and assesses our action in terms of these values. Practical values should not be confused with other values, aesthetic or moral. Whether we are aware of this or not, it is through constructing praxiological models that we accomplish progress in technology. Progress means an improvement of the measure of effectiveness in at least one aspect. Usually the praxiological model assumes some losses in effectiveness in order to attain more substantial gains. It is sometimes fascinating to analyze how meticulous and impeccable is the calculus of gains and losses in the praxiological model, which very often is constructed without an awareness of its praxiological nature.

It seems to me that if the characterization of technological progress as the pursuit of effectiveness is correct, the philosopher of technology must include the study of praxiology and in particular the study of praxiological models in his inquiry. Organization theory is simply inadequate for this purpose because of its limited scope. The advances of modern technology take on a very complex form requiring integration of a variety of heterogeneous factors as well as the establishment of a hierarchy of levels. What finally matters is the increased measure of effectiveness, but the road to this increase is multichanneled and multileveled. Traditional organization theories are unable to handle this complexity, but praxiology can.

Technological progress, analyzed in terms of measures of effectiveness, led us to two questions. The first was whether technological effectiveness can be treated in general terms – this prompted us to consider praxiology. The second was whether we can distinguish specific patterns of thinking leading to the increase of effectiveness in different branches of technology.

I shall devote the remaining part of this paper to the second question. That is, I shall attempt to discern specific patterns of technological thinking for some branches of technology. I do not propose to find such patterns for technology as a whole. What I can offer are some suggestions as to how one may approach the problem and discern these patterns in less complex fields. If the procedure is right, it will lead to the discovery of other patterns in other branches of technology.

Before I attempt to spell out some of the structures or patterns of thinking in technology, I shall show what they are and how they work in microbiology. The microbiologist makes daily observations of microscopic sections which are quite simple from a certain point of view. Now what is a microscopic section, for example, of a diphtheria culture? It is, in the layman's language, a specific configuration of certain forms which possess characteristic structures. This is how far we can go in describing the phenomenon verbally. In other words, no amount of verbal explanation will render it possible for the layman and generally for the untrained person to recognize the diphtheria culture by mere description. At first the layman and beginning students of microbiology are simply unable to perceive what is there to be seen. After some period of training they do perceive and are in fair agreement as to what they see. The ability to recognize certain microscopic structures is thus peculiar to students of microbiology.

The art of observation is not universal but specific for a given field or subject matter. When ever observation plays a significant role in scientific investigation, it is selective observation directed toward perceiving some objects and their configurations and toward neglecting others. Observation, however, is not only a perceptual process but also involves some conceptual thinking. Certain types of observation are intrinsically connected with thinking in terms of certain categories.

In general, it seems to me that *specific branches of learning originate and condition specific modes of thinking, develop and adhere to categories through which they can best express their content and by means of which they can further progress.* I shall illustrate this thesis by examining some branches of technology, namely, surveying, civil engineering, mechanical engineering, and architecture, with the understanding that the last, architecture, is only in part a branch of technology.

I will start with surveying. The final products of surveying are maps, plans, and profiles in elevation. In order to avoid complications in the analysis, instead of considering a map that is a projection of a larger area of land on a sphere, I shall examine a plan that consists of a projection of a smaller piece of land on a plane as the referential surface. It is quite obvious that we must measure all angles of the figure to be projected on the plane, all its sides, and at least one azimuth. Now, the specific questions for this surveying operation, and indeed for all geodesy, are: Why is this method applied, not any other? Why should we measure the sides with a metal tapeline and not by steps or by eye? Why should we check and adjust our instruments? A surveyor, who is quite capable of skilfully performing all the geodetical operations, might be less capable of relating all these operations to one theme, one central element that accounts for the specificity of surveying. It is one thing to follow a procedure and another to be able to grasp and verbalize the essence of this procedure or, in other words, to make measurements and to be aware of the specific structures of thinking characteristic for surveying.

What, then, is specific for thinking in surveying? It is *the accuracy of the measurement*. This can be seen while tracing the development of surveying from its earliest stages as well as while following its recent progress. In the final analysis, it is always the accuracy that lies at the bottom of all other considerations. Sometimes it is expressed in an indirect and disguised form, for example, when we inquire which of two or three methods is most economic or most efficient. However, even in this case, the silent assumption is that the accuracy remains the same or, at any rate, that the decrease of accuracy is negligible and the economic gains – which sometimes may be of prime importance – are quite considerable. It is thus the most conspicuous feature of geodesy that it aims at a progressively higher accuracy of measurement; in an indirect form this may mean a reduction of cost or time or work while preserving the same accuracy. Thus, we may say in a succinct form: *To think geodetically is to think in terms of accuracy.*

Succinct forms have the virtue that they pin down one crucial element of the analysis; they have the vice that (for the sake of brevity) they neglect other elements and consequently present a simplification of the phenomenon under investigation. So it is with our succinct characterization of geodetical thinking. It is by no means the only kind of thinking the surveyor performs. It is not even the dominant thinking in terms of the actual time devoted to it. But thinking in terms of accuracy is the most *instrumental* for surveying. And that means that the practitioner of surveying will be a better practitioner if he is aware of the specificity of geodesy and if he applies consistently his knowledge in his practice. And this also means that the researcher in geodesy will be a better one if he consistently keeps in mind that geodesy aims at a progressively higher accuracy of measurement. Furthermore, the grasp of the specificity of surveying will help the scholar who investigates the history of surveying. History of any branch of learning is twisted and full of unexpected turns and blind alleys. Unless we discover the 'Ariadne's thread' in its development, a history of any discipline will be but a mosaic of unrelated or loosely related events, descriptions, theories. Thus, the discernment of patterns specific for a given branch of learning is not only an activity that may give us the comfort and aesthetic satisfaction that accompanies neat classifications for the sake of classification but may indeed be of a concrete value to the practitioner, researcher, and historian. It is in these terms that I deem the analysis of patterns of thinking important.

To return to technology, when we consider a typical civil engineering project – whether the construction of a house or a bridge – the decisive element is the *durability of the construction*. Therefore, we may say that *thinking, specific for the civil engineer, is in terms of durability*. Durability is the starting point, or at any rate the ultimate element of the analysis. The choice of materials and the methods of construction must be related to the required durability.

Theoretical research in civil engineering is directed toward the discovery of combinations of materials that will either increase the durability (of the construction) or lower the costs at the same durability. During the execution of a project, some calculations may be made and the accuracy of the calculations taken into account, but here they are of subsidiary importance. The main issue is durability, although admittedly the form of its manifestation may be very complex or disguised.

Perhaps this can be seen even more clearly when we review the history of civil engineering or, in other words, when we review the history of architecture in its constructional aspect. If we omit the aesthetic and utilitarian aspects, the history of architecture can be seen as the development and perfection of those architectural forms and those combinations of materials that increase durability. Although the progression of more and more durable forms is often hidden under the guise of artistic trends and movements, it is there and can be traced easily.

Turning now to architecture proper, architectural thinking is simultaneous thinking in terms of durability and aesthetics and utility, and the two latter categories are perhaps more important than the first one. When projecting a house, the civil engineer must consider new materials and their combinations as well as new constructional designs. When designing the same house, the architect must consider the standards of comfort, hygiene, and, generally speaking, the 'livability' prevailing for his times, as well as the aesthetic tastes of his epoch, its predilections and aversions. Thus, thinking in terms of utility and artistic predilections separates the architect from the civil engineer.

I shall now very briefly consider mechanical engineering. The key element in this branch of engineering is efficiency (in the narrow sense of the term when it refers to the efficiency of an engine, whether steam or combustion). Thus, *thinking, specific for mechanical engineering, is in terms of efficiency* (efficiency here is meant in the narrow sense specified above). In designing engines, the problem of efficiency has two aspects: either we attempt to increase the absolute efficiency and raise it as close as possible to 1, or we attempt to construct a 'better' engine while keeping the same efficiency ('better' can mean: safer, cheaper, longer lasting, more resistant). Obviously, certain problems concerned with the strength of materials have to be considered and solved, and therefore thinking in terms of durability takes place here as well; it is, however, of a derivative character. By saying it is derivative, I do not mean to say that it has little significance or no significance at all but, rather, that the starting point for an analysis of durability are problems of efficiency. Problems of durability are not chosen at random but are selected with an eye to the solution of the problems of efficiency.

In considering machine tools, the question of efficiency is not immediately obvious but may be shown to be of crucial importance as well. A number of other factors, such as the cost of construction, durability, and useful life, are analyzed at the same time. Finally, we either attempt to raise efficiency while preserving the same cost, the same useful life, and the same durability; or we attempt to reduce the cost while preserving the same efficiency, the same useful life, and the same durability; or to prolong the useful life with the remaining data unchanged.

To summarize, to think in terms specific for a given discipline is to think in those terms that (a) determine the lines of investigation within this discipline; (b) account for the historical development of this discipline; (c) explain the recent growth of the discipline.

Once again it should be emphasized that categories specific for various branches of technology or, more generally, specific for various branches of learning, are not those that end all but rather those that begin all. They are the key to the analysis. They are the key to the idea of technological progress. Neither should it be surmised that categories I call specific have anything to do with Kantian categories. Perhaps my terminology is unfortunate. My point was simply to draw attention to certain patterns of thinking which can be discerned as characteristic for various branches of technology and elsewhere. The most important conceptual elements in these patterns I call categories.

I should not be surprised if the 'categorical' analysis as sketched here will be viewed as insufficient for an exhaustive epistemological description of technology. Perhaps it should be remembered that as yet no general philosophy of science – which after all has been developed for some centuries – is viewed as sufficient. Can we then expect more from a subject that is beginning to emerge than we expect from a related subject that has achieved a considerable maturity?

The analysis of the structure of thinking in technology is hampered by the fact that nowadays the construction of bridges, highways, automobiles, or even domestic gadgets is inseparably linked with the consideration of beauty and comfort which are basically 'non-technical' categories. Technical categories, such as accuracy and durability, are, so to say, the technological constants. They are the yardstick of technological progress. Aesthetic satisfation and comfort are to a certain degree variables. They cannot be measured objectively for all epochs. The more decisive their influence on the object designed, the more difficult it is to recapture the structure of thinking peculiar to a given branch of technology. Architecture again can serve as an example.

Luigi Nervi, Oscar Niemeyer, and Frank Lloyd Wright, among others, are architects from whom the element of a construction (e.g., the beam of a house) is often at the same time a component of an over-all aesthetic pattern. These constructor-architects think at the same time in terms of durability and in terms of aesthetic satisfaction; they find aesthetic expression in functional, that is, purely constructional elements. A similar situation occurs in other domains of technology. While designing and constructing automobiles or lathes, can-openers or intercontinental ballistic missiles, the purely technical aspects often are interwoven with aesthetic and utilitarian aspects. The technological phenomenon no longer is identical with the technical phenomenon and cannot be analyzed entirely in terms of the engineering sciences. The social context, the economic structure of a society, the existing social mores and aesthetic predilections – all have their imprint on the technological phenomenon, and to a certain extent, determine its character.

In summary, I should like to observe that mistaken ideas about the nature of technology reflect what technology was a century or two centuries ago and not what it is today. In the twentieth century, and particularly in our day, technology has emancipated itself into a semi-autonomous cognitive domain. There are many links between science and technology, but a system of interrelations should not be mistaken for a complete dependence. A fruitful way of reconstructing the epistemological status of technology is through grasping the idea of technological progress. Technological progress is the pursuit of effectiveness in producing objects of a given kind. The purely technical elements, such as the accuracy or durability or our products, are often considered in larger economic frameworks which complicate the basically technological typology and even impede the analysis in terms of purely technological categories. In addition, the standards of beauty and utility are becoming intrinsic ingredients of technological products, and this makes our analysis even more difficult. However, our task is to meet these difficulties, not to avoid them. The point is that the structure of technology is far more complex than the methodologist of science is prepared to admit. It is only through recognizing this complexity, and through granting to technology a methodological autonomy, that we may be able to end the stagnation in a field which as yet has only a name – the philosophy of technology.

Los Angeles, Calif.

NOTES

* First published in *Technology and Culture* 7 (1966), 371–383.

[1] By the 'technological object' I mean every artifact produced by man to serve a function, it may be a supersonic airplane as well as a can-opener.

[2] *Praxiology – An Introduction to the Science of Efficient Action*, London, 1965. See also my article, 'Praxiology – the Science of Accomplished Acting', *Personalist*, Summer 1965.

I. C. JARVIE

THE SOCIAL CHARACTER OF TECHNOLOGICAL PROBLEMS*

Comments on Skolimowski's Paper

There is a great deal in Professor Skolimowski's very interesting paper with which I fully agree, especially, perhaps, his thesis that technology is not to be identified with science and that it has a different philosophy and methodology. If there is a single problem that is central to Skolimowski's paper it is perhaps his statement that science has a single aim, the pursuit of knowledge or truth; can a similar, simple aim be ascribed to technology? Skolimowski's answer is 'yes' and that that aim is efficiency, the construction of ever more efficient solutions to technological problems. So while scientific progress is toward truth, technological progress is an increase in efficiency.

In the last half of his paper Skolimowski pursues the question of in what exactly efficiency consists. His view, roughly, is that it is different in different branches of technology. Different patterns of thinking lead to efficiency in different technological subjects. These differences of patterns of thinking he outlines with respect to surveying, civil engineering, and mechanical engineering. Efficiency in surveying consists largely in the pursuit of accuracy; in civil engineering it consists in durability, in mechanical engineering it consists in 'efficiency' in the technical sense of the word. Skolimowsky expresses the hope that from these beginnings patterns can be discerned for technology as a whole, and indeed that this might even lead to 'a new methodology of science'.

Technology seems to have been treated like Cinderella by philosophers of science. It has always been put in the second-best place, mentioned almost as an afterthought. This is perhaps understandable, since the received notion of the role of technology is that it is the province of engineers and other such non-gentlemen, and its philosophy thus is not a matter of great concern to the philosophical purist. This attitude of the philosophers is especially baffling to the present-day public who expect the philosophy of science to discuss such achievements as the atom bomb, space rockets, supersonic airplanes, television, and so on. Whenever one of these examples is introduced in a discussion philosophers of science are tempted to say: "But that is *technology*, of no *scientific* importance at all."

The philosophers of science have so far disdained technology despite the fact that we are living in an age of technology and not an age of science. Whereas ancient Greece and the seventeenth and eighteenth centuries were the eras of science, with discussion and progress going on apace, today we have had, for example, an impasse in quantum physics for a generation. Technology, on the other hand, grows and progresses all around us. Little new *science* is involved in a manned landing on the moon, but a great deal of new technology is. In setting 1970 as a target date for the accomplishment of this feat, NASA's technologists boldly proclaim that they expect to be able to make enough technological progress before then – to iron out the wrinkles, as they say – to carry the program through. Can one imagine someone setting a target date for the completion of a project that involved a unified field theory? How then to approach this neglected subject of a philosophy of technology?

Skolimowsky very correctly began with an extensive review of what technology is or, rather, what it is not. He then passed to two problems: first, that of distinguishing technology from science (he did this with reference to the idea of progress in both); second, that of isolating the characteristics of specifically technological thinking.

I would like to break the philosophy of technology into three main philosophical problems: (1) In what ways does technological knowledge grow and progress? (2) What is the epistemological status of technological statements? (3) How are technological statements to be demarcated from scientific statements? These problems overlap, of course, and perhaps are not of equal importance, but they seem to me to pretty well cover the main initial questions any philosophy of technology must answer.

I

To begin with, then, how does technology progress? My own field of competence is social anthropology, and hence I shall refer to primitive societies in dealing this problem.[1] It is, I hope, obvious enough that all societies have to cope with technological problems. Their ability to solve them obviously will be a function of the clarity with which they are posed, the resources and experience available, and, as Agassi has very rightly emphasized,[2] luck and ingenuity.

Societies do not face the same set of technological problems of course; African tribes are not faced with the problems involved in going to the moon or with the problem of coping with excessive traffic in cities. These are cases where highly developed technology has created, so to speak, its own further technological problems. Yet all societies face the common problems of food, shelter, and transport, to take only three examples. Skolimowski touched on shelter when he discussed house-building and on transport when he discussed bridge-building. He did not touch on food. One might think that before the era of canned and frozen foods, food was not a technological problem. But of course domestic animals like cows, pigs, and hens – and especially battery hens and crops like high-yield wheat and seedless oranges – are as much technological achievements as Metrecal. In the field of transport the carefully bred horse is as much an invention as the motorcar or the airplane; in the matter of shelter, the mud hut is a forerunner of the skyscraper. I find quite unintelligible the idea that technology means only machines, especially in an age when the dividing line between animate and inanimate matter is increasingly hard to draw.

So in a sense all societies have technology of sorts, and the problem of progress involves comparing them with each other in as unprejudiced a manner as possible. Not all primitive technology has been superseded. There are parts of this great country where only a horse can take one – the horse, as has been said of the bicycle, has reached its own plateau of evolution; there are times and places where a shack or even an igloo is as effective a shelter as a modern house; and of course we have by no means progressed beyond the food inventions of earlier generations.

Not all technological progress is replacement and abandonment of previous means. New means used to solve old problems, old means used to solve new problems, and new means used to solve new problems are all easy to illustrate. The old problem of flying was solved with the new means of the airplane; the new problem of high-speed calculation was solved with the old means of the circuit, the switch, and binary numbers, that is, the latest computers (although perhaps the combination was new, but certainly the new problem of going to the moon is being tackled with the old, old means of the rocket); and finally the new problem of traffic congestion was solved by the new means of freeways. Technology seems to me to progress in all these ways.

But is this a specific enough treatment of how technology progresses? Can we say no more than that progress is a function of the clarity, resources, and ingenuity that can be brought to bear on a technological problem? Is there no heuristic special to technology? Skolimowski seems to be moving in this direction when he posits that the general aim of efficiency and the means of achieving it in different branches of technology is to give emphasis to particular qualities, like accuracy, or durability, or 'efficiency' in its technical sense.

Some arguments occur to me which indicate that efficiency is by no means the aim always underlying technological progress. Take Skolimowski's suggestion that efficiency in civil engineering primarily consists in thinking in terms of *durability*. Is durability necessarily an aim of building? Does it not depend on what the building is for? Blockhouses of World War II were very highly durable, and as a result they still scar parts of the European landscape – because of the expense of removing them. Americans *could* build fireproof, floodproof, typhoonproof, and earthquake-proof houses, but they are discouraged by the expense. They are also discouraged by the fact that they cannot make them progress proof. They cannot, that is, build houses that will not be out of date sooner or later – and these days it is sooner rather than later. Built-in obsolescence is as important a consideration in houses as in motorcars. Similarly, I think one could argue that in surveying, *accuracy* is only a relative aim, relative, that is, to the problem the surveyor and his cartographer are set. Terrain maps may ignore land use, road maps and tenancy plans may ignore terrain, maps of the London and New York subways ignore all but the most general features of what they are depicting.

What these criticisms move toward is something like this: it is the problem that is posed to the technologist which determines the character of the thinking required or, as I would prefer to say, the overriding aim that is to govern the solution. The engineers who devised the Bailey Bridge and the pontoon bridge were given civil engineering problems in which *speed of construction* was far more important than *durability*. A map-maker may be far more concerned with a clear general outline than with particular details. The designers of American cars certainly are not much concerned with the efficiency of their product so much as the appearance, the smoothness of ride, etc. American cars could be much more durable and much more efficient (in terms of cost per mile) were manufacturers prepared to raise the unit cost. But the unit cost must always be considered alongside the problems of worker or consumer satisfaction, aesthetic attraction, and social cost. Whether the overriding concern is with accuracy, durability, 'efficiency', or what, is always dictated by the socially set problem and not the technological field. Here, then, I take issue with Skolimowski when he says there are patterns of thinking in technology specific to the branch and independent of the sociological background.

The very fact that we live in an age of great technological progress in itself suggests we are not in dire need of a heuristic gift from philosophers, always provided that we can give some explanation of this technological success, an explanation to the effect that a heuristic is already in use or one is not needed. My own suggestion is that the success is explained by the increasing clarity with which technological problems are posed, by our improved ability to think ahead. Less and less are inventions made and then found uses for; more and more are they sought when there is a demand or a potential demand. The element of inventiveness is still very great, but all that can be done in the way of a heuristic has been done. This tremendous concentration on single problems tends to accelerate progress, since not only solutions but multiple solutions may be found, and then *these* may turn out to have other applications.

II

I come next to the problem of the epistemological status of technological statements. I am puzzled as to why Skolimowski contrasts science as the exploration of the real, with technology as the creation of the artificial. It seems to me that the constrast is less between the artificial and the real as between the artificial and the natural. Science and technology are both concerned with a world that is equally real, but the latter is concerned to interfere with it. Science, of course, aims to investigate the whole universe and to describe the most general laws it obeys. Technology is quite different; it works within what these laws allow, concentrating on what is possible in narrow localities of the universe and definitely not everywhere. For example, much of our technology must be changed when we enter weightless or low-gravity environments, just as big Tokyo buildings are different from big New York buildings on account of earthquakes. In terms of the logical analysis of the statements of science, then, technological statements are part of the factual description of conditions in a circumscribed locality of the universe, what are called the 'initial conditions'.[3]

And, of course, very often we can discover initial conditions, such as the fact that a certain drug cures a certain illness, without having any explanatory theory of why they are so, just as fish swim without hydrodynamics and primitives grow crops with no scientific theory of how this occurs. Here we seem to have some explanation of why there are societies that have technology but nothing we would recognize as science; we hardly find the reverse situation (since there are no societies without technology). This leads me to dissent from Skolimowsky's view that technology is not concerned with the pursuit of truth, although I would concede that successful technology sometimes can be in conflict with accepted science. For example, ships and space ships navigate with tables based on Newtonian celestial mechanics, which is part of Newton's physics – and Newton's physics is superseded. Thus truth cannot be the principal aim of technology. However, technological information is not unrelated to truth. The fact that the stresses to which the fuselage structures of high-speed aircraft are subject result in a phenomenon called 'metal fatigue' was an important discovery about the world, and our airplane technology had to take it into account. Like gravity it is a very localized phenomenon. A scientific explanation of the fact of metal fatigue was not a necessary prerequisite of a technological solution to it.

III

This brings me to the last of the three problems mentioned above, namely, the demarcation between scientific statements and technological statements. Agassi has discussed this, and Skolimowski's views are not particularly dissimilar. We all agree that the aims and methods of technology are different from those of science and that truth is not the overriding aim of technology. Wherein, then, lies the core of the differences? The answer is, I am sure, that there is no absolute demarcation:

it is very much a matter of context, and particularly of problem context. Scientific statements are posed to solve scientific problems; technological statements allow that certain devices are not impossible.

When a problem is or is not scientific can hardly be formally characterized, but in a general way science must involve questions about the nature of the world. Which problems are in fact studied seems to depend upon the scientific tradition and also upon the empirical character of the theories from which the problem arose.

Similarly, no formal criterion of technological problems can be expected. It is again a question of the aim. If the aim is to manage to accomplish something practical which science allows as not impossible, rather than to discover some universal truth, then technology is what we call it. Both science and technology are, as Skolimowski pointed out, goal-directed and thus amenable to praxiological analysis. The question remains whether our agreement that the structure of technology is complex and its methods diverse can be maintained in the face of our disagreement over whether there are specific ways of thinking peculiar to branches of technology independent of a social and traditional context.[4]

Toronto

NOTES

* First published in *Technology and Culture* 7 (1966), 384–390.

[1] Although in anthropology I am *against* functionalism (the doctrine that things ought to be explained solely in terms of their social function) and instrumentalism (the doctrine that scientific theories make no claim to truth, only to being useful instruments), in the philosophy of technology, following Skolimowski, I am *for* both.

[2] See Agassi, 'The Confusion between Science and Technology in the Standard Philosophies of Science' [see pp. 40–59 of this book].

[3] In the model of scientific explanation proposed by Karl Popper in 1934 (see his *Logik der Forschung*, Vienna, 1934 (transl. as *The Logic of Scientific Discovery*, New York, 1959), pp. 59–60) a statement is explained if it can be validly deduced from premises containing one or more universal laws (UC), together with one or more particular statements of fact – the initial conditions (IC). An example of this would be: (UL) All men die when their heads are cut off. (IC_1) King Charles I of England was a man. (IC_2) King Charles I of England had his head cut off. Therefore, King Charles I of England died. For extensive discussion see C. G. Hempel, *Aspects of Scientific Explanation*, Glencoe, Ill., 1965, *passim*; and W. W. Bartley, III, 'Achilles, the Tortoise and Explanations in Science and History', *British Journal for the Philosophy of Science* **13** (1962), 15–33.

[4] The author is indebted to Professor Joseph Agassi for helpful discussions in the preparation of these comments.

F. RAPP

TECHNOLOGY AND NATURAL SCIENCE – A METHODOLOGICAL INVESTIGATION*

I. NATURE AND ARTIFACTS

As a first step in an attempt to delimit natural sciences and technology it would seem sensible to consider the original meaning of the two words. The term 'natural sciences' would thus denote the sciences concerned directly with nature, whereas technology – in accordance with the Greek *techne*[1] – would have to do with objects produced artificially by man. On the one hand there would be nature, left to itself and untouched by man, as investigated by the natural sciences, and on the other the world of technology as an artificial creation which is only brought about by intervention in this natural order.

Strictly speaking, a further distinction ought to be made between the original condition of inanimate nature and the changes wrought on it solely by plants and animals, without human assistance. But such changes are, for the most part, not considered to constitute intervention. They are regarded as a modification inherent in a comprehensive order embracing both animate and inanimate nature. From our present-day point of view, the interventions made by primitive man and by the tool cultures of earlier epochs also appear as comparatively slight modifications of a 'natural' course of things that remains in principle unchanged.[2] In contrast to these the large-scale, systematic interventions in the natural process conducted by modern technology are unmistakable. Here the 'artificial' character of all technical procedures and products becomes clearly apparent.

As regards the manner in which they come to exist, there is in principle no difference between technical constructions and works of art; both are artifacts. Nevertheless, a distinction can be made on the basis of their different functions: technical artifacts always serve a *practical* utilitarian purpose. They relieve man of physical labour – and, in the case of computers, from routine mental operations as well –, by making certain natural processes serve his goals. The artifacts created with artistic intent, on the other hand, are designed from the start with a view to *aesthetic*

F. Rapp (ed.), Contributions to a Philosophy of Technology, 93–114. All Rights Reserved

contemplation. Manifested in them is a specifically artistic idea, which finds expression in the way the object is fashioned, and is intended to evoke a particular effect in the beholder.[3]

At this point the fluid transition from technical constructions to works of art becomes apparent. In many cases a change in the attitude of the beholder suffices to make a technical artifact into a work of art. Indeed, as something constructed by man, deliberately and with a particular sense of form, every technical object is really at the same time implicitly a work of art as well. For this reason modern technology tries to combine practical function and artistic form by appropriate design. Looked at like this, the reverse procedure, whereby modern art uses technical processes and objects as means of expression, appears entirely consistent, albeit the artistic content of such endeavours remains in many cases dubious.

Less convincing, on the other hand, is the attempt to delimit the natural sciences from technology by assigning to them as their field of activity nature, untouched by human hand. In the teleological view of nature in Aristotelian physics, according to which every natural movement is directed towards an end peculiar to it,[4] every change brought about by man did indeed appear as intervention in the natural course of things. But in the view of modern natural science there is in principle no difference between natural events, which occur of themselves, and phenomena caused by man. On the contrary, the great advance of modern natural science was only rendered possible by abandoning this distinction.[5]

Only a diminishing fraction of what can today be regarded as indisputable in the findings of research in the natural sciences can be perceived by means of direct observation of untouched nature. The fundamental results of modern natural science were only attained through a combination of idealizing mathematical theory and systematically conducted experiments. The phenomena that are to be investigated by means of a particular theoretical approach have generally first to be created 'artificially' in appropriate experimental conditions. The necessary observations, too, can in most cases only be conducted with the aid of complicated measuring instruments.

The fact that the various phenomena investigated are not simply present, but have first to be induced on the basis of a particular theory and with the aid of complex apparatus and instruments, has given rise to the various constructivist and operational interpretations of natural science.[6] It must be borne in mind, however, that a totally arbitrary approach is not possible, either with regard to theoretical conceptions or to active intervention in nature. Thus, although from a purely logical point of view it is possible to undertake any heuristic projects at all, provided only that they are consistent in themselves, scientific research can only use theories whose empirical consequences do indeed ensue in appropriate experiments. Furthermore, the phenomena induced 'artificially' do not differ *fundamentally* from 'natural' processes occurring spontaneously. There is a constant flux between the two: in particular it must be borne in mind that many natural processes can also be induced experimentally. Furthermore, even the most complicated indirect measurements must always ultimately be rendered as direct perceptions of the senses in some way if they are to become accessible to our experience at all. And, finally, it also becomes clear from the interaction of various theories and the unexpected support they often receive from a more distant context, that the various theoretical conceptions, with the aid of the appropriate experiments, really do lead to a grasp of 'objective reality'.

Consideration of the notion of nature and artifacts as two separate, opposed spheres thus leads one to the conclusion that if one argues from the way in which the artificially produced technical objects come into existence, then they must indeed be termed artifacts, and as such distinct from what is encountered in untouched nature; but, on the other hand, it is not possible to limit the scope of research in natural science to nature as untouched by man. For in the modern natural sciences the phenomena to be investigated must first be isolated, as it were, with the aid of complicated technical apparatus and instruments, so that in an abstract sense they too can be regarded as artifacts. The distinction between nature and artifacts is thus not a suitable basis for a fundamental differentiation between natural science and technology. Both cases involve deliberate intervention in untouched nature, resulting in the production of particular objects or phenomena which are basically subject to the same laws and are therefore in principle capable of being investigated by the same procedures.

Thus, neither for the research worker in natural science nor for the engineer is there a fundamental distinction between natural and artificially induced phenomena. As the objects of scientific research and technical activity the two are regarded as being in principle of the same kind. This situation is so familiar to anyone who has even only a fleeting acquaintance with natural science and technology that it really requires no special mention and only appears worthy of note in connection with the attempt to set up a clear antithesis between the natural and the artificial. But it is only because they are of the same kind that one can describe with the same concepts and the same mathematical formulae both the relations existing in untouched nature and the processes induced by man. This is why, on the one hand, the results of research in natural science can be applied in the production of technical artifacts, and, on the other, the phenomena artificially induced with the aid of complicated technical apparatus and observed by means of highly sensitive instruments afford a sort of continuation and refinement of the experience familiar from direct sensory observation.

From what has been said so far it follows that, strictly speaking, one ought really to distinguish two concepts of nature: (1) nature in the narrower sense of the natural course of things, *without human interference*, as distinct from all the artifacts created in scientific research or technical production, and (2) *nature in the wider sense*, embracing equally the natural course of things *and* everything produced artificially. Nature in this all-embracing sense is the object of attention for both natural science and technology. For natural science it constitutes the object of research and for technology the substratum of the implements produced by man and the processes he initiates. The laws governing it are *essentially* the same for mere contemplation, active intervention and technical inventions. Of course, this does not exclude the treatment, for *pragmatical* reasons, of other processes and the use of other concepts in the forming of theories in natural science, which are intended to contribute to a comprehensive systematic description, than in the technological sciences, which are designed for practical application.

Whereas the first, narrower concept of nature certainly has a concrete, sensory content, the second, wider concept can really only be understood as an abstraction, for here the accent shifts from the observable phenomena to the material *substratum* which is the basis of the various manifestations.[7] This abstract character becomes still more apparent when one takes into consideration the fact that the object investigated imposes no definite rules for the forming of theories and the articulation of natural laws. One has to find fruitful modes of questioning and promising research projects, which, moreover, always remain tied to a historically contingent conception of scientific inquiry. Nevertheless, nature, which is the basis of *both* technology *and* natural science is not only abstractly conceivable, for the empirical testing of theoretical approaches always guarantees a link with the world perceptible to the senses: the abstract, material substratum is constantly manifesting itself in concrete form.

As what already exists in nature and what is produced artificially are reduced to the same level, the world is made to appear no longer as an inviolable cosmos to be accepted with reverence, in which man himself has his assigned place, but as an object merely present, at his arbitrary disposal. That nature can be seen in another way, too, is not only apparent from a retrospect of human thought. A vague sense of it still survives today in the artistic and cosmically embracing attitude to nature, the effect of which, unconsciously and in a rudimentary form, extends to the conception of nature as a sphere of leisure and recreation.

It is in this context that the destruction of the natural environment and the rape of raw material resources must be seen. Because the equilibrium established by the natural order in a long process of gradual development no longer has precedence, everything that is technically possible can in principle be carried out in practice. As soon as nature is regarded only as available raw material, as an object to be changed, then this consequence is indeed inevitable. For the sober, enlightened mind, the material world offers no 'resistance' – except for the laws to which it is itself subject – to any kind of utilization to which it may be put. The natural forces are thus, so to speak, blind to the context in which they work; they can be guided in any direction we impose on them. Since nature sets us no limits here, we must impose the necessary restrictions on ourselves. Responsibility towards coming generations thus constrains us to pursue our exploitation of natural forces only within sensible bounds.

II. THE INTERCONNECTION BETWEEN TECHNOLOGY AND NATURAL SCIENCE

If one considers the mutual dependence of technology and natural science, two tendencies can be discerned in our present situation, which can be epitomized as the increasing *assimilation* of technology to natural science and vice versa. The first is apparent in the adoption by technology of findings and methods of research in natural science. Thus, for instance, the *findings* of physical investigations of the high vacuum and thermal emission of electrons led to the construction of the amplifier tube. Only by this means did present-day radio, radar and television technology become possible,[8] though the electron tubes have recently been replaced by transistors, which are derived from solid state physics. Similarly, automation and data-processing are dependent on basic research. In other fields, such as nuclear and laser technology, the physical origin of the technical applications is directly discernible.

In addition to new principles, which may then, in certain cases, lead to the development of entire branches of industry, scientific research supplies technology with better materials, more efficient production processes and improved tools and machines.[9] It can be taken as a rule of thumb that sooner or later scientific findings are applied in technological practice in some way or other. As a result of the powerful pressure of competition in the economic and armament spheres the time lapse in this process between discovery of new findings and technological utilization of them is being constantly reduced. At least as important, however, as the revolutionary upheavals that consist in the opening up of entirely new possibilities are the detail improvements in existing processes, for in general new principles only become capable of economic utilization after continued improvement of details.

In addition to research findings technology is also taking over the *working methods* of natural science. Today the place of overall *ad hoc* trials, based on actual practice, which, in the case of complex phenomena, at first constitute the only possible approach, is increasingly being taken by the forming of theories, based on mathematical description and on systematically conducted experiments. This tendency is apparent in the individual, relatively isolated special fields of technology, just as in the more comprehensive approaches of cybernetics and systems engineering or in the attempts to establish a theory of design.

In *basic research* there is no distinct dividing line between technological and scientific investigations, so that clear assignment to one or other of the two fields is often no longer possible. Here research not directed at specific goals but only intended to extend theoretical understanding, as customary in natural science, becomes the guiding principle. There is at first no thought of practical application of the results attained, although on the basis of past experience it is rightly supposed that most of the knowledge obtained in these basic investigations can sooner or later be put to some technical use.

Big science,[10] organized as teamwork and deliberately planned, also belongs in this context. Its task is, on the one hand, to find scientifically based principles for solutions to given *technical problems*, and, on the other, to open up possible spheres of technical application for *scientific research findings* already obtained.[11] By means of such planned inventions many new products have been evolved and developed to the production stage, particularly in the chemical and electronics industries.

Just as scientific knowledge and methods are entering technology, scientific research is in turn determined by technology. Thus the assimilation of technology to science has its counterpart in the dependence of science on technology. Two ways can be distinguished in which technology exercises an influence on science: firstly, it provides instruments and apparatus for scientific investigations, and, secondly, technological development throws up new fundamental problems that stimulate the course of scientific research.

A brief glance into a modern laboratory is sufficient to show the extent to which the range of *instruments* available to the scientist is determined by technology. Without the intricate measuring instruments and apparatus, which range from the simple Geiger counter via amplifying systems and vacuum equipment to electron microscopes, wind tunnels and particle accelerators, today's scientific research is inconceivable. Such technical equipment is also indispensable for maintaining the constancy of experimental conditions and transmitting and amplifying the values measured. Even the evaluation of observational data obtained is frequently only possible with the aid of computer technology.

Indeed, many objects of scientific research, such as, for instance, the transuranic elements, many isotopes and elementary particles and most organic compounds are only created at all through the use of technical aids.[12]

As a kind of counterpart to the technologist's scientific equipment, this technical apparatus belongs to the range of research instruments the scientist takes for granted. Naturally, by means of a broad enough definition all these items of technical apparatus could also be counted as part of the sphere of natural science, since they are based on the application of scientific principles and have in many cases been constructed for the specific requirements of scientific research. But as material constructions, produced by man to fulfil a particular function, they are to be regarded – just like, for instance, a turbine or a car – as technical products. If this distinction were to be abandoned, the *difference* between natural science and technology would be obliterated completely. On the other hand, however, it would also be a mistake to regard the natural sciences, on account of their technical instruments, as a part or a mere by-product of technology;[13] for in this case the types of approach and methods of procedure peculiar to natural science would not receive due attention.

In addition to this direct dependence of natural science on technology there is the indirect influence, which stems from the occupation of science with *tasks* arising within the field of technology and indicating new directions of scientific research. For instance, the question of the deformation of metals, so significant for materials engineering stimulated investigations on plastic flow and dislocation in crystals.[14] Further examples are the study of superheated steam at high pressure, the investigation of fatigue and tracking current phenomena, the corrosion of materials, the structure of technical catalyzers, the composition of organic compounds and the aerodynamic drag in supersonic flow.[15] In all such cases technical application engenders a demand for further basic scientific investigation. As a consequence, the direction taken by scientific research is determined not solely by immanently scientific considerations but also by the problems thrown up by technology – especially on account of the financial means provided for the improvement of technical processes.

Despite this close and growing interconnection between natural science and technology, it must be remembered that the two fields continue to remain *relatively autonomous*. Thus particularly in the investigation of atomic dimensions and in the development of comprehensive theories the natural sciences have their own field of activity, where independent of technological questions the systematic investigation of natural phenomena is conducted primarily for its own sake. And the technologist, for his part, is certainly not always dependent on natural science for help in solving his concrete problems. Frequently he has to construct what is required just on the basis of the – possibly incomplete – knowledge available in the particular case, and without precise acquaintance with the natural phenomena involved. Since he cannot wait for the results of thorough, detailed investigations, he makes do with the knowledge available, as long as this knowledge can provide a technically viable solution. Consequently, the practice of engineering is in many cases ahead of scientific research.

On account of this autonomy of technical development, it would thus be wrong to regard technology simply as applied science.[16] This applies to the *present state* of technology and science. But as regards the field of human knowledge to which they belong, all the processes applied in technology are based on the exploitation of natural phenomena. For this reason, technical processes always constitute a *potential* object of research for natural science. And, vice versa, it is always *possible* – though certainly not in every case economically viable – to apply the results of scientific research in technological practice.

Among the fields of technology in which overall *ad hoc* experiments are used, without recourse to systematic, theoretically-based research, are, for instance, the combustion problems in the combustion chambers of rockets, where evaporation, mixing, combustion and dissociation ensue in so small a volume and within so short a time that detailed scientific analysis is hardly possible.[17] Further areas where the method of pragmatical experiment dominates and, even without knowledge of the physical bases, has led to success, are problems of friction, rheology, fracture and deformation, processes of flow and reaction in the various motors, and chemical catalysis.[18]

If technology is to be split up into two large fields, each *dominated* either by the application of theoretically-based scientific knowledge or by practical experience acquired experimentally – though with the complementary aspect still playing a significant role in each – the division that emerges is roughly as follows: Practice is the decisive factor in the cases of the technology of transport, metallurgy, mining and building. Fields primarily dependent on basic research include electronics and modern communications technology, production engineering, measuring and control technology as the basis of the mechanization and automation of processes, and also computer technology and medical physics.[19]

Altogether, then, the present relationship between technology and natural science offers a picture – differentiated according to the various aspects and special fields – of manifold connections and instances of mutual dependence. For this reason it would be inappropriate to reduce the complex interconnections that in fact exist to a unified pattern. Particularly inaccurate are all extreme interpretations whereby, depending on which aspect is stressed, either the natural sciences appear as a by-product of tasks set by technology, or technology is regarded as mere application of the results of scientific research.

On the other hand, one cannot be content either simply to draw the two fields together, on account of their close interconnection and treat them as constituting an entity, a single complex of science-and-technology. As soon as attention is shifted from the borderline field of basic research to specifically technological or scientific projects, clear distinctions can be observed. For instance, it is obvious that working out a unified field theory and constructing a motor car certainly constitute problems of different kinds. Thus treating science and technology as a single complex would in any case have necessitated *immanent* differentiation and involved the question of delimiting the two again.

III. VARIOUS AIMS

What then is really the root of the difference that evidently exists between technology and natural science? It would seem probable that more light could be thrown on the question by examining the two fields in greater detail than in the relatively summary treatment they have received so far. The complex of *natural science* can be split, roughly speaking, into two components: the *research process* and the *results* arrived at thereby. From the methodological standpoint, a further distinction can be made between the goal of attaining knowledge, which is the purpose of all scientific investigation, and the structure of the specifically scientific methods of research, which are adjusted according to the nature of the particular phenomenon being investigated. Viewed as a whole, then, natural science constitutes a methodical and consequent process of research, yielding reliable knowledge of the laws governing the material world.

The complex of *technology*, on the other hand, is considerably more comprehensive and, at the same time, only very vaguely differentiated. But at any rate a rough distinction can be made along the lines of the following four points: (1) the command and application of acquired skills and utilitarian procedures; (2) a group of academic disciplines that provide the engineer with the knowledge necessary for the production and use of certain material objects; (3) the production process itself in its methodological structure and as concrete activity, stretching from the initial planning to the final production; and (4) the objects thus produced, their functions and their use.[20]

In view of these heterogeneous components, which in turn are all composed of various different elements, a comprehensive, unified definition of the concept is only possible if certain delimitations are made within the given range of interpretation. This being so, in the context of the present discussion technology can be summarily characterized as the harnessing of nature for human aims, certain material objects being produced on the basis of suitable knowledge and skills in a preconceived, consistently conducted process of production, and then put to use in an appropriate manner.

This characterization of the two fields shows that natural science and technology, strictly speaking, do not overlap all along the line, but merely at one point. In actual fact, natural science, as a whole, and the technological sciences, as academic disciplines (see (2) above) enter into a state of competition, as they are *both* concerned with scientific investigation of the material world. This means that the counterpart to natural science as a *whole* consists of just *one part* of technology. And it is only with regard to this special aspect, where technology appears as a scientific enterprise, that overlapping may occur. Furthermore, the close interconnection between the two fields also originates in this sphere of common concern.

This would seem to imply that discussion of the relationship between technology and natural science should have been confined to comparison of the natural and technological *sciences*. But whereas the natural sciences constitute a field complete in itself, the technological sciences are so closely linked with the whole *complex of technology* that isolated treatment of them would not appear to be a very fruitful proposition. Technology is always ultimately concerned with the exploitation of particular laws of the material world for the production and use of corresponding implements and apparatus. And the technological sciences are really only a *means* serving this practical purpose. For this reason, a meaningful comparison becomes possible only if the process of production and the objects produced are also taken into consideration.

One could of course object that research in natural science fulfils an instrumental function as well. The research process is not conducted for its own sake, but with the aim of acquiring precise, comprehensive and critically examined knowledge about the structure of the material world. In accordance with the classical scientific ideal of *theoria* as disinterested mental contemplation,[21] the sole purpose of research in natural science is to acquire understanding of natural laws and phenomena. The orientation of the scientist's pursuit of knowledge is fundamentally theoretical. Any possible application of the results of his research is in general a matter of indifference to him; he seeks knowledge for its own sake.

This contemplative, theoretical attitude does not, however, preclude the possibility that the knowledge acquired may only come through active, practical intervention in the natural process. As became apparent in the analysis of the relationship between nature and artifacts, the most important discoveries of modern natural science have been based on experimental investigations in which the phenomena to be investigated have first had to be 'artificially' isolated and induced by means of suitable technical apparatus. But in natural science this active intervention is always subordinate to the aim of forming theories. And the theories themselves are intended to describe the material world, not to facilitate practical alteration of it. So the active intervention concerns only the *genesis*, not the *application* of the theory.

These statements apply, however, only to the *intention* behind research in natural science. The *factual content* of the knowledge thus acquired and the *possibilities* of applying it are certainly not exhaustively characterized thereby. For with regard to systematic procedure – i.e. without going into the historical development – it has to be noted that the reason why within a research project in natural science the phenomena to be investigated are isolated and induced experimentally is that the tentative theories have to be tested to see whether they hold good. Only by systematic experimental control can sound theories be distinguished from arbitrary speculations. A theory can be said to have come through an empirical text successfully if it has been possible to derive from it certain predictions which are consequently fulfilled. From the point of view of logic, a *prognosis* like this is largely identical with a *scientific explanation*: both are concerned – from a formal point of view – with deriving principles concerning individual events from more general laws by means of appropriate specification.[22]

Of decisive significance for the systematic connection between technology and natural science is the fact that the same predictions can be used in technology – merely by changing the direction of interest – to solve practical problems. Supposing, for instance, that from a well-established theory the following prognosis can be derived: if the state (or process) *A* comes about, then according to a natural law the state (or process) *B* will follow. The *test* of this prediction will consist in inducing *A* by suitable measures and seeing whether *B* does in fact ensue. What is tested in this case is thus whether the anticipated *connection* between *A* and *B* does really exist. Because thereby *ipso facto B* is induced, the same procedure – based on the same prediction – is also effective if the starting-point is the *technological task* of producing the *state* (*or process*) *B*.

From a methodological point of view, the predictions that can be derived from a known natural law are independent of the functions they are to fulfil. In scientific research they are used for testing the tentative theory, and in technology for solving practical problems. This distinction between *testing* and *application* is the reason why the realization of technical projects is based only on proven theories, whereas in scientific research the examination of predictions deduced from heuristic hypotheses not yet verified is one of the principal aims.[23]

Thus, whereas in natural science the primary object of interest is theoretically oriented acquisition of knowledge, in technology practical application always occupies the foreground. In a very general sense, therefore, one can say that the relation between natural science and technology is one of *theory and practice*. But this description is only partially correct and subject to reservations. In as far as they constitute an insight into the pattern of natural laws, from which predictions can be deduced, all scientific research findings *can* – in principle – be put to technological use. But they will only be applied in some actual technical project if in addition materials and procedures are known which permit the project to be carried out on the required scale without costs exceeding an economically acceptable figure. Thus, of the scientific knowledge under consideration in each case, only that portion which fulfils these requirements is put to technological use.

So in the fields of technology particularly dependent on scientific research findings the task of specifically *application-oriented technological research* consists in fulfilling the technological requirements for successful utilization of the findings. In this field of technological research concerned with practical application, where it is often necessary to work with *ad hoc* suppositions introduced pragmatically if the aim of the research is to be attained at all, theoretical reflections naturally play a part *as well*. Results of investigations of this kind enter into the body of systematic knowledge in the technological sciences. To sum up, then, we can say that on the one hand not all the findings of scientific research are in fact put to use in technological practice and on the other hand technological practice has its own body of theory, which is not directly dependent on natural science.

The place in society occupied by the two fields under consideration has not so far been treated in this analysis. In as far as they involve actions technology and natural science only exist as a part of the life of social activity, but this point of view seems to escape attention in a methodological investigation that concentrates entirely on the immanent logic of the procedures used in the two scientific fields. Thus the process and the results of *scientific research* are determined to a large extent in advance by the aim of a mathematical description as comprehensive and precise as possible and by the experimental method to be used. Influences from 'outside', which are alien to the discipline, would appear to be of only secondary significance.

True, here too critical examination does reveal quite a number of influential factors which are not cogently determined by the object of the particular investigation. Both the problem itself and the heuristic outline of a solution to it are not quite simply given, in isolation, but depend on the prevailing scientific views, the state of research, the financial resources available and, finally, on the personal style of the researcher as well. Furthermore, in the practical conduct of research the criteria according to which decisions are made concerning the confirmation or refutation of a theory always operate within a certain latitude of discretion. Despite these reservations, however, the structure and results of scientific research – at least within a particular period of scientific history – are relatively independent of extra-scientific determinants.

With *technological* problems the situation is fundamentally different. Whereas the aim of scientific research is to investigate a specific field of phenomena without the manner in which the knowledge acquired is subsequently to be formulated being rigidly determined beforehand, technology is ultimately always concerned with seeking solutions to specific problems, in connection with which it is confronted from the start with a catalogue of demands. Whereas the scientist engaged on a research project is prepared to be surprised by his findings, the technologist has to seek solutions to his problems within a strictly limited latitude. In technology, then, the knowledge used in each particular case, the production process in which particular technical objects are manufactured and, finally, the application of them, are always subordinate to a (specific social) aim which constitutes a *prior* determining factor for the purely technical procedures. Methodological analysis thus shows that at this point one is necessarily led beyond the technical sphere to social determinants.

Here the difference in the aims becomes clear once again. For scientific research there is – subject to the reservations already made – a kind of immanent self-determination, because in the natural sciences *cognition* of the structure of the laws governing the material world is an ultimate aim, not subordinate to any other task. In contrast to this, technology is always concerned with *concrete realization*[24] of material objects which are to fulfil particular functions and in so doing must satisfy prior demands. These demands are determined, for instance, by the needs to be fulfilled by a particular technical construction and by the price a potential user is prepared to pay for it.

In the technological sciences concerned with general investigations in connection with the structure and production of possible technical constructions, such prior determining factors are not *explicitly* in evidence. They always, however, play a part *implicitly*, for it is according to them that in practice-oriented research the area is delimited within which the investigation of technical possibilities seems to have any point at all. However, as soon as the precise specification for a technical project is laid down, the various prior determining factors come into explicit effect.

IV. DIFFERENCES IN PROCEDURE

The differing aims of technology and natural science also result in different methods of procedure. To obtain a serviceable technical construction or an acceptable scientific theory, it is always necessary to carry out specific operations the individual stages of which have to follow a certain predetermined pattern. The relevant methodological principles, however, do not concern the actual activity, which, for instance, can be regarded with respect to the persons acting and made the subject of research in psychology and sociology. What is of interest here is just the general rules of procedure, as conditioned by the structure of the field concerned, according to which the various purposive actions are performed.[25]

If a *scientific research process* is considered in a simplifying way, two stages can be distinguished in it which could be designated summarily as the phases of conception and testing. In the stage of *conception* the task consists first in taking a heuristic model and on this basis proposing certain working hypotheses concerning the structure of the phenomena to be investigated. These hypotheses are then formulated by means of general mathematical equations from which – ideally – other, hitherto unknown phenomena can be deduced besides those observed so far.

During the stage of *testing*, specific predictions have to be deduced from the general laws, concerning the observational data which can be expected in the particular case under investigation. After these theoretical operations the empirical testing proper consists in inducing the phenomena to be investigated, by means of appropriate experimental equipment, and establishing the values of the variables concerned. The hypothesis under investigation is corroborated by such a test if the values obtained by experiment are sufficiently close to those calculated theoretically.[26]

Similarly, every *technical production process* – broadly speaking – can be divided into a phase of conception and one of realization. In the stage of *conception* the general principles of construction are first established, on the basis of the demands which the desired technical object is to fulfil. This initial concept enables increasingly specific designs to be obtained with the aid of preliminary tests, culminating in a final plan of construction. In the phase of *realization* the actions implicitly prescribed by the rules for construction are in fact carried out, so that at the end of the process the desired material object is produced. All complicated cases require in addition special methodological techniques, such as, for instance, specialization and division of labour, the use of suitable tools and equipment, the manufacture of standardized units or assembly-line production.

Since technology *and* natural science involve purposive processes of action, there must be a phase of conception in both cases. Nonetheless, this stage has a different significance for each field. In the scientific research process it is used to determine the general hypotheses, whereas in the technical production process the plans for construction are worked out. Whereas in natural science we find formation of *hypotheses*, followed by *deduction*, technology is concerned with the realization of a *projected* object in a *pragmatically* oriented production process. In contrast to the *hypothetico-deductive* method of the natural sciences one could perhaps speak of the '*projective-pragmatical*' method of technology.

It must be borne in mind that scientific research too, of course, which is a systematic and purposive process of action, must be based on a specific concept and, where appropriate, on a particular plan as well. Over and above this real actions are necessary, in order to make the logical inferences and carry out the experimental investigations which are required. Something similar applies to the technical production process too, which, for its part, involves forming hypotheses and drawing deductive conclusions. Thus the summary terms suggested do not give an exhaustive account of all the components present. Each of them only emphasizes the aspect which dominates in view of the particular aim – the forming of theories or the production of concrete objects.

The differing aims and procedures of technology and natural science explain why one and the same field – namely the pattern of laws governing the material world – can be subjected both to technological and scientific investigations, which, although they are interconnected and overlap at many points, belong to different spheres of scientific study. This is because, in order to express in terms of laws the structure of the material world, which itself does not impose any particular formulation, it is necessary in each case to extract from the abundance of phenomena certain characteristics and relations, on which all further investigations are then concentrated.

In the historical development of science this selection from among the competing axioms and conceptual models has been made with a view to obtaining simple, comprehensive and empirically corroborated theories. Thus despite the common historical origins of science and technology in the Renaissance, in the course of time the specific theoretical framework we encounter in present-day natural science has emerged as a distinct view of nature.[27] And the conceptions which have proved especially suitable for practical exploitation of nature, have constituted the bases of technology.[28]

Thus it is the particular aim of natural science to establish theories which are as comprehensive as possible and which – ideally – yield predictions of any desired degree of accuracy for all relevant phenomena. To attain this aim *idealizing* assumptions are made, which constitute only an approximation to observable phenomena, but are particularly suitable as a basis for mathematical description. These approximations, such as, for instance, the supposition of point particles or completely elastic bodies, can then be adapted to the given conditions by means of appropriate corrections. This tendency to *abstract* from the phenomena directly perceptible to the senses, which is characteristic of natural science, becomes especially evident when purely abstract mathematical concepts like tensors and symmetry-groups are used. In summary, one could thus speak here of the process of *idealizing abstraction.*

In contrast to the wide-ranging deductive theories of natural science, designed to yield mathematical description which is as precise as possible, technology is content with relatively isolated findings, which, however, are related from the start to specific tasks in the construction or use of technical objects. For this reason the concepts applied in technology are generally 'closer' to reality,[29] for the real aim consists here in obtaining sufficiently precise values for the relevant parameters concerning, for instance, the dimensions of a technical construction or the efficiency of a particular process.[30]

Harnessing the forces of nature for human aims in the production and use of technical objects was described as a process of action in which a preconceived plan is realized ('projective-pragmatical' method). This term only applies to the general *scheme of action* which is valid for *all* planned and purposive techniques, even if they do not concern man's exploitation of nature. With respect to physical reality, however, any 'technique' consists in the final analysis in inducing specific natural processes in as pure a form as possible. The technical objects produced during a purposive process of action are designed precisely in order to *realize* such desired natural processes. To obtain optimal efficiency, the aim is to eliminate as far as possible all other more or less inevitable secondary effects, such as, for instance, losses by friction. All processes are *concentrated* towards obtaining the one desired effect.[31] '*Realizing concentration*' could be used as a summary term for this.

The peculiar *overlapping* of science and technology can be explained by the differing procedures, designed in the case of science to facilitate comprehensive mathematical description, and in the case of technology to enable material objects to be produced in which certain physical processes are concentrated. Due to the universality of the approach of natural science, all the physical processes occurring in technology can, in principle, be subjected to scientific investigation. But certainly not all the processes in fact used in technology can be derived in all their details from known laws of nature.

In order to induce such technical processes as are based on particular scientific principles – like jet propulsion on the law of momentum – appropriate material devices are necessary whereby the desired process can indeed be brought about on the scale required and with the necessary efficiency. The maximum that can be attained in technical practice is determined by the conservation laws of natural science.[32] From the technological standpoint it is always concrete realization and not theoretical description that constitutes the decisive problem. For this reason a scientific *discovery* can in general only be applied successfully on the basis of corresponding technical *inventions*. Because the processes used in technology are frequently the result of *complex interaction* between manifold physical laws, problems of basic research in technology generally require interdisciplinary cooperation. To the practical engineer, however, such detailed investigations of complex processes are only of interest if they yield an improvement in technical efficiency.

Hitherto it has been tacitly assumed that science and technology can to a large extent be regarded as *homogeneous* fields which do not require further subdivision. But this assumption is not justified in every respect. Thus in natural science, according to the thesis of T. S. Kuhn, the difference between '*normal*' and '*revolutionary science*' has to be taken into account. In the case of normal science an accepted scientific theory, such as classical mechanics, is elaborated in detail. Revolutionary science, on the other hand, consists in a fundamental change in the mode of scientific thought, an example of which is the transition from classical mechanics to quantum mechanics.[33]

An *analogous distinction* can be made in technology. Within a particular technical field detailed improvement of a given and – in principle – unquestioned process constitutes the counterpart of normal science. For, just as in the case of science, what happens here is that a procedure the principle of which is taken for granted is constantly being developed. Thus the treatment of a scientific problem within the framework of a particular research project has its counterpart in the technological development work by which a procedure is constantly being improved and refined. The fundamental technological innovations – corresponding to revolutionary science – consist then in the transition to a completely new principle, whereby hitherto unknown possibilities of application are opened up. But just as in natural science the fruitfulness of a new approach can only be assessed on the basis of its concrete elaboration, so in technology the efficiency of a new procedure can only be judged definitively when it has been developed in all its details.

Berlin

NOTES

* The German version of this paper was published under the title 'Technik und Naturwissenschaften – Eine methodologische Untersuchung', in *Techne – Technik – Technologie (Philosophische Perspektiven)* (ed. by H. Lenk and S. Moser), Pullach 1973.

[1] Cf. J. E. Heyde, 'Zur Geschichte des Wortes Technik', *Humanismus und Technik* **9** (1963), 25–43.

[2] H. Rumpf, 'Wissenschaft und Technik', in *Die Philosophie und die Wissenschaften (Simon Moser zum 65. Geburtstag)* (ed. by E. Oldemeyer), Meisenheim 1967, pp. 89–91.

[3] Cf. H. Pogoda, 'Technik und Natur, Technik und Kunst', *Deutsche Zeitschrift für Philosophie* **18** (1970), 57–76.

[4] Aristotle: *De caelo* I 4, 217a33; *De anima* II 4, 415b16.

[5] The historical development will not be the subject of further consideration here. For details see among others E. A. Burtt, *The Metaphysical Foundations of Modern Physical Science*, London 1932; and A. Koyré, *Études Galiléennes*, Vols. I–III, Paris 1939.

[6] H. Poincaré, *The Foundations of Science* (transl. from the French), Lancaster, Pa., 1913; P. Duhem, *The Aim and Structure of Physical Theory* (transl. from the French), New York 1962; H. Dingler, *Das Experiment-Sein Wesen und seine Geschichte*, Munich 1928; P. W. Bridgeman, *The Logic of Modern Physics*, New York 1927. Recent publications include K. Holzkamp, *Wissenschaft als Handlung*, Berlin 1968; P. Lorenzen, 'Wie ist die Objektivität der Physik möglich?' in *Argumentationen-Festschrift für Josef König* (ed. by H. Delius and G. Patzig), Göttingen 1964, pp. 143–150.

[7] Cf. E. McMullin (ed.), *The Concept of Matter*, Notre Dame, Ind., 1963.

[8] E. Appleton, 'Partnership of Science and Engineering', *The Engineer* **198** (1954) 853; cf. also J. R. Bright, *Research, Development and Technological Innovation*, Illinois, Ill., 1964, p. 71ff.

[9] O. A. Saunders, 'The Scientist's Contribution to Mechanical Engineering', *The Engineer* **210** (1960), 760.

[10] Cf. A. M. Weinberg, *Reflections on Big Science*, Cambridge, Mass., 1967.

[11] The systematic search for new possibilities is especially emphasized in K. Hübner,

'Von der Intentionalität der modernen Technik', *Sprache im technischen Zeitalter* **25** (1968), 27–48.

[12] J. Müller, 'Zum Verhältnis von Naturwissenschaft und Technik', in *Klassifizierung und Gegenstandsbestimmung der Geowissenschaften und der technischen Wissenschaften* (Freiberger Forschungshefte D 53), Leipzig 1967, p. 165f.

[13] Thus J. K. Finch says: "There are, however, some thoughtful scientists who feel that science has become a mere appendage of technology and are disturbed by the widespread use of the term 'science' or 'scientific' for many interests, activities, and products which have little or no connection with truly scientific pursuits." ('Engineering and Science', *Technology and Culture* **2** (1961), 330).

[14] H. Rumpf: *op. cit.*, p. 102.

[15] Cf. 'Engineers and Scientists' (Leading Article), *The Engineer* **197** (1954), 715.

[16] Cf. M. Bunge, 'Technology as Applied Science', *Technology and Culture* **7** (1966), 329–347; and H. Skolimowski, 'The Structure of Thinking in Technology', *ibid.*, p. 372. [see pp. 19–39 and p. 73 of this volume].

[17] O. A. Saunders, *op. cit.*, p. 760.

[18] H. Rumpf, *op. cit.*, p. 102.

[19] R. Rompe, 'Grundlagenforschung und Technik', in *Sitzungsberichte der Deutschen Akademie der Wissenschaften zu Berlin* (Klasse für Mathematik, Physik und Technik), **1967**, p. 35.

[20] The delimitation of the German concepts 'Technik' and 'Technologie' is discussed by H. Lenk, *Philosophie im technologischen Zeitalter*, Stuttgart 1971, pp. 133–135, and by S. Moser, 'Technologie und Technokratie – Zur Wissenschaftstheorie der Technik', in *Neue Aspekte der Wissenschaftstheorie* (ed. by H. Lenk), Braunschweig 1971, pp. 171ff.

[21] Aristotle, *Nicomachean Ethics* X 7, 1177 a11–1177b4; similarly Plato, *Timaeus* 89d–90d.

[22] Further details can be found in C. G. Hempel, *Aspects of Scientific Explanation*, New York 1965, Ch. 4; E. Nagel, *The Structure of Science*, London 1961, Ch. 3; W. Stegmüller, *Probleme und Resultate der Wissenschaftstheorie und analytischen Philosophie*, vol. I, Berlin 1969, Ch. 1.

[23] Cf. J. Agassi, 'The Confusion between Science and Technology in the Standard Philosophies of Science', in *Technology and Culture* **7** (1966) 364. [see p. 55 of this volume]. The significance of falsification of scientific hypotheses is emphasized especially by K. R. Popper, *The Logic of Scientific Discovery*, London 1959.

[24] Cf. K. Tuchel, 'Zum Verhältnis von Kybernetik, Wissenschaft und Technik', in *Akten des 14. Internationalen Kongresses für Philosophie*, vol. II, Vienna 1968, p. 582. He says "die Materialität ihres Bestandes scheint mir ein wesentliches Kriterium der technischen Gebilde zu sein."

[25] F. Rapp, 'Die Technik in wissenschaftstheoretischer Sicht', in *Neue Aspekte der Wissenschaftstheorie* (ed. by H. Lenk), Braunschweig 1971, p. 180.

[26] M. Bunge, *Scientific Research*, vol. I/II, Berlin 1967; E. Nagel, *op. cit.*; K. R. Popper, *op. cit.*; W. Stegmüller, *op. cit.*, among others, undertake investigations of the methodological problems of the natural sciences.

[27] Historical change in scientific thinking has been discussed in recent years by S. Toulmin, *Foresight and Understanding*, London 1961; T. S. Kuhn, *The Structure of Scientific Revolutions*, Chicago 1962; I. Lakatos and A. Musgrave (eds.), *Criticism and the Growth of Knowledge*, Cambridge 1970; G. H. von Wright, *Explanation and Understanding*, London 1971.

[28] The significance of the differing criteria of progress for the delimitation of technology and science is emphasized by H. Skolimowski, *op. cit.*, pp. 371–385.
[29] Cf. A. Hermann, 'Die historische Relativität naturwissenschaftlicher Erkenntnis', *n + m (Naturwissenschaft und Medizin)* **6** (1969), 3–10.
[30] E. Jobst, 'Spezifische Merkmale der technischen Wissenschaft in ihrem Wechselverhältnis zur Naturwissenschaft', *Deutsche Zeitschrift für Philosophie* **16** (1968), 934 mentions the diagram in particular as being an aide typical of the engineer [see p. 131 of this volume].
[31] J. Müller speaks of the technologist going beyond the reality he is confronted with by concentrating ('verdichten') particular properties and relations (*op. cit.*, p. 164).
[32] The rôle of the limiting possibilities is stressed by L. Tondl, 'Slovo o filosofii techniky', *Filosofický časopis* **12** (1964), 281–293 [see pp. 1–18 of this volume].
[33] T. S. Kuhn, *op. cit.*, esp. Ch. 3 and Ch. 9.

M. I. MANTELL

SCIENTIFIC METHOD – A TRIAD*

I. INTRODUCTION

The 'scientific method' may be defined as a time-tested broad pattern of problem solving which gives efficient results. Many individuals such as John Dewey, have claimed that the scientific method can be used for all problem solving. Others argue that there are many types of problems for which the scientific method cannot be used, particularly for complex human relationships such as those found in politics and sociology. This paper will follow the thesis that the scientific method may generally be applied to all types of problems; and that where the method appears inapplicable it appears thus because of application of the wrong pattern of the scientific method's triad of patterns of problem solving.

It has been recognized that there are significant differences between the pattern of problem solving generally used by engineers and others in the applied sciences, and the patterns of problem solving used by researchers in the basic sciences. However, relatively little attention appears to have been paid in the literature to the possibility that there are not just two, but three, distinct patterns of problem solving, all of which fit into the broad sequence of the 'scientific method'.

These three patterns within the scientific method, in order to cover the widest range of problems, may be entitled: (1) Basic Research; (2) Applied Research; and (3) Systems Approach. Each pattern is applicable only to a characteristic type of problem; efficiency in solving each characteristic type of problem would be anticipated to improve by a better understanding of the general pattern involved. A knowledge of the significant difference between the three patterns also should aid in understanding the differences between the normal work assignments of the engineer, the scientist and the technician, and also understanding of terms such as 'development', 'operations research', and 'systems analysis'. Each of the three patterns will be presented in subsequent sections.

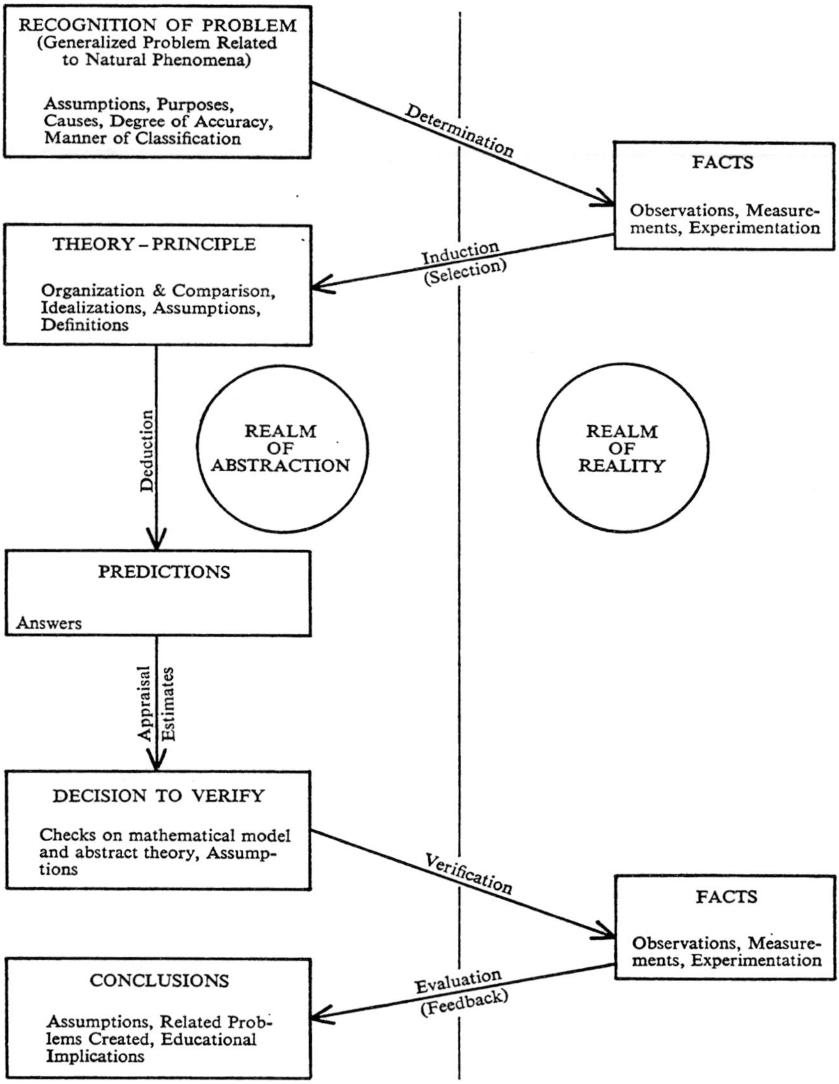

Fig. 1. Scientific method (basic research).

II. BASIC RESEARCH

Figure 1 shows a flow chart of the sequence of steps involved in an efficient approach to basic research. This pattern of problem solving would characteristically be applied to generalized problems related to natural phenomena, that is, seeking general theories or principles applicable to all of a species, similar elements of matter, etc. Recognition of the problem, and many of the other steps in the sequence, take place in the realm of abstraction of the mind; while other steps involve observations, measurements, and possibly experimentation, in the realm of the real world (reality). A key step in the sequence is the process of induction in which a theory or principle is derived or inferred from the initial set of specific observations and measurements. The induced theory may only by coincidence agree with the initial observations and measurements (such coincidence is the origin of most superstitions); and therefore, a verification process is vital for an efficient solution to the problem.

For any specific set of observations and measurements, a number of theories may be induced; and for a particular problem one of these possible theories must be selected for verification. The verification process is used on a set of predictions obtained by application of the selected new theory or principle. The predictions are obtained by the process of deduction in which a specific fact or conclusion is derived or inferred from a general theory or principle – the solution of a mathematical model for a specific value would be the most common form of deduction. A decision to verify would be preceded by appraisal and estimates of the reasonableness of the predictions and the involvements required in the verification process. Final evaluation (and feedback) of the facts observed in the verification process (measured against the predictions) lead to conclusions as to whether or not the new theory or principle has been verified or disverified, whether related problems have been created, and the education implications as to reporting the results. Evaluation and feedback, of course, may take place at any point in the sequence, which may then involve a modification of the original problem or of one of the previous steps.

III. APPLIED RESEARCH

Figure 2 shows a flow chart, following the general sequence of the 'scientific

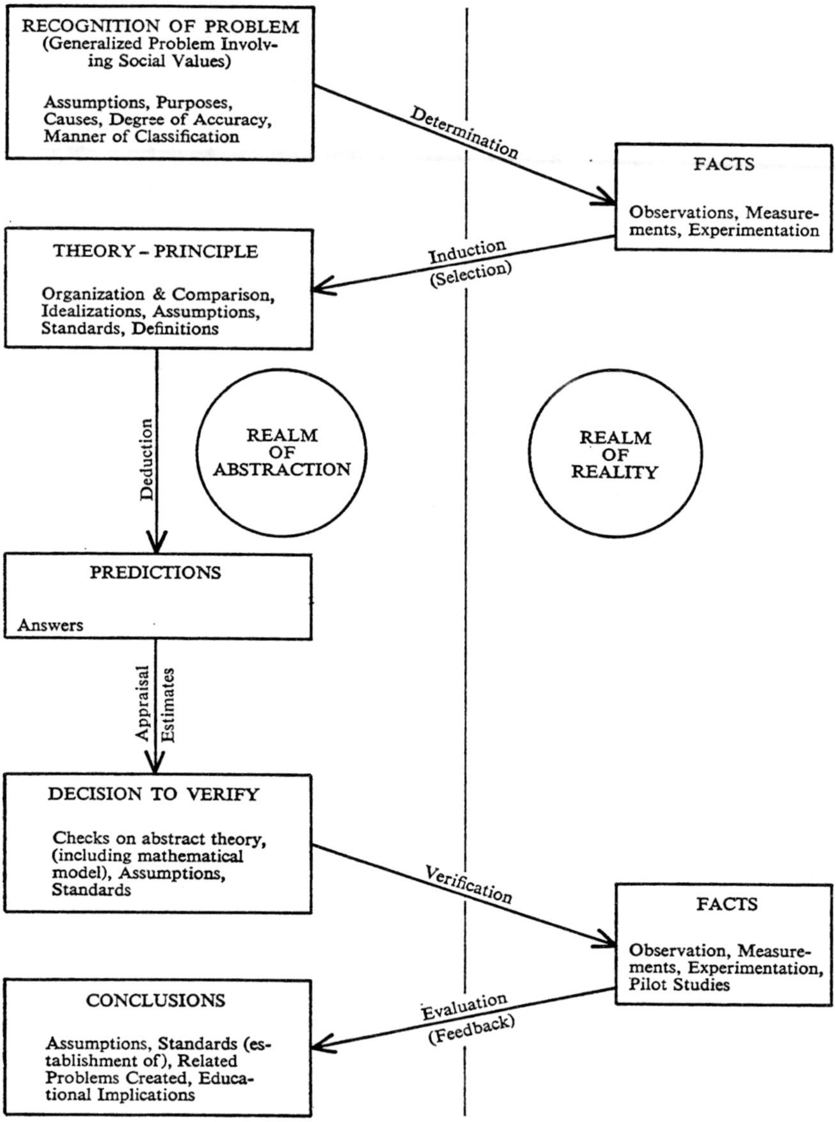

Fig. 2. Scientific method (applied research).

method', for the pattern of problem solving used in applied research. This pattern would characteristically be used for generalized problem situations seeking theories or principles related to materials, procedures, systems, etc., having social values or social applications. In basic research, knowledge is sought only for knowledge's sake – the characteristics of a species of butterfly may be studied without, necessarily, considering its utility to man as a food or other product; a study of the charge on the electron may be made without any thought of developing an atom bomb. Applied research would more readily accept less accuracy in describing phenomena if simplicity in application were achieved.

The most significant difference between the patterns of problem solving in basic and applied research is in the types of assumptions used. In both basic and applied research a typical assumption might be the close approximation to the ideal state, such as the weightless member or frictionless environment (when weight and friction are negligible). However, only in applied research (and the practice thereof) are there assumptions or standards which involve elaborately considered value judgments requiring compromises between demands or safety, cost, durability, reliability, efficiency, convenience, aesthetics, etc. Typically, a committee of experts studies and recommends realistic bounds and constants as standards (e.g. factor of safety) which are approved or disapproved at a national convention.

This prior social agreement on basic assumptions (standards) resulting in codes and specifications is a major factor behind the success of engineering practitioners. Similarly, a major factor behind the advances in medical practice is the use of applied research to establish standards for dosages, circumstances for application, procedures, etc., of new drugs and methods of treatment. If each individual engineering or medical practitioner were to use only his own value judgments for each problem, the relatively consistent success presently achieved would be unlikely. In the practice of endeavors such as politics and sociology, where most of the arguments claiming inapplicability of the scientific method originate, there apparently has been little or no effort to agree on basic assumptions (standards) involving the complex value judgments needed for the related applied research.

IV. SYSTEMS APPROACH

Figure 3 shows a flow chart, following the general sequence of the 'scien-

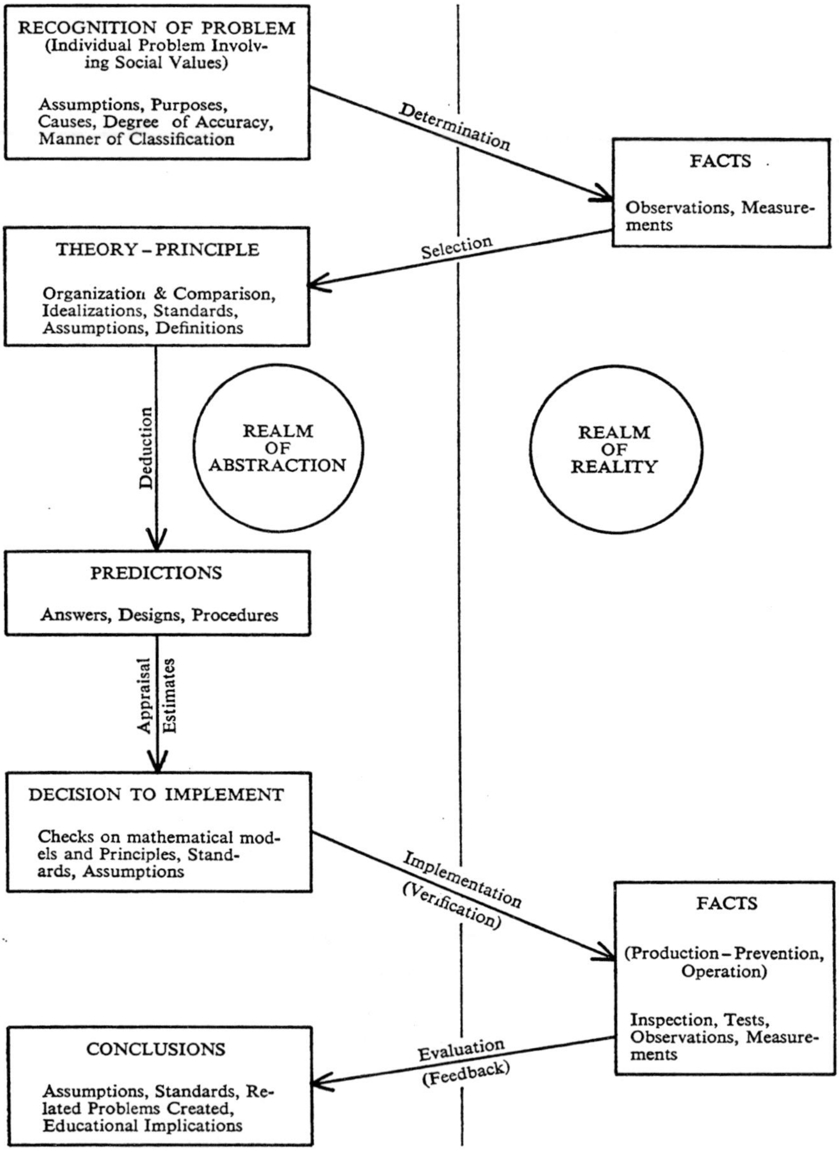

Fig. 3. Scientific method (systems approach).

tific method', for the pattern characteristically used in solving individual problems directly applicable to man's needs and desires. This is the pattern generally used by engineering, medical and other practitioners. The major differences between the three patterns of problem solving are summarized in Table I.

TABLE I

Scientific method – component elements in patterns of problem solving

Component elements (1)	Basic research (2)	Applied research (3)	Systems approach (4)
Generalized problem situation	×	×	
Individual problem situation			×
Reference to social values		×	×
Observation, measurement	×	×	×
Experimentation	×	×	
Inductive process – establishing theories	×	×	
Selection of theories – principles	×	×	×
Deductive process	×	×	×
Assumptions, as standards		×	×
Prediction, as an aim	×	×	
Appraisals – estimates	×	×	×
Verification, as an aim	×	×	
Implementation, as an aim			×
Evaluation and feedback	×	×	×

The first major difference from the previous patterns is the lack of an inductive process establishing new theories or principles, and the selection of one of these. Instead, the practitioner selects the appropriate existing and previously accepted theories, principles and standards he considers applicable to the problem. Next, in the deductive process predictions are made not as an aim for verifying, but to obtain answers, designs, and procedures which can be implemented. Implementation results in a form of verification, when implementation is successful, but verification is not an aim in this pattern of problem solving. Implementation typically involves a production or prevention process with inspection taking place during and tests taking place after the process. Subsequently, operation and maintainence of the system would typically be necessary.

The specialized 'functions' of design, production, inspection, operation, etc., may be noted as steps in the total sequence of solving a major problem. These 'functions' will be found in each problem, regardless of the 'field', such as Civil, Electrical or Mechanical Engineering, in which the problem may be classified. The 'development' function is essentially a repetitive application to the same problem of the latter process (the systems approach), with major attention to feedback. Training of technicians is typically pointed for functions performed in the 'realm of reality' and also for simple deductive processes based upon standards selected by others.

Numerous subproblems, including the above mentioned functions, must be solved for a large project (problem); and increased efficiency would also be desired in the solution of each subproblem. 'Operations Research' appears to have first developed as a pattern of 'applied research' attempting to establish new general theories applicable to improving (the function of) operations in the military field during World War II. Subsequently, practitioners have applied these same theories to other individual types of operations and have, incorrectly, continued to refer to these applications as operations 'research' even though no research or new theories are involved.

The term 'systems approach' has been used herewith in recognition of the fact that all problems involve some form of system, e.g. structural system, electrical system, man as a system. No single system is independent of the infinite system, referred to as the universe; however, the experienced practitioner is able to isolate those elements of the infinite system which will have the major effects upon the problem at hand. The term 'systems approach' is sometimes, incorrectly, applied only to instances in which systems are enlarged somewhat to include factors which were previously ignored – if these factors were previously unjustifiably ignored. then the previous problem solution may still be a systems approach, but with an inadequate system. The term 'systems analysis' has come into frequent usage, primarily referring to analysis of the limited area of production or operation systems. It has been recognized that most operations research work is not actually research, and also there has been a current tendency for operations research to limit its endeavors to merely describing the operations problems by complex mathematical models, without getting into the implementation stage. 'Systems analysis' has therefore been used to connote a more broadly encompassing term than 'operations research', but much more restrictive than the 'systems approach'. Some also distinguish between 'operations research' and 'systems analysis' by assigning the former title to operations with 'existing' hardware and the latter to operations with 'new' or futuristic hardware.

V. SUMMARY

There clearly appear to be three distinct patterns of problem solving falling within the broad sequence of the 'scientific method': (1) Basic Research; (2) Applied Research; and (3) Systems Approach. The engineer, the physician and other practitioners are directly concerned, primarily, with the last two of these patterns of problem solving. One of the most significant differences between the work of the basic scientists and the practitioners such as the engineer is in the types of assumptions used, where the successful practitioner uses basic assumptions (standards) involving a complex system of prior social agreements resulting in codes and specifications. Conversely, the apparent general lack of success in attempted solutions of social and political problems may largely be attributed to the failure by the practitioners therein undertaking to obtain prior social agreement on the basic assumptions involved.

Coral Gables, Fla.

NOTE

* First published in *Proceedings of the American Society of Civil Engineers* **95**, No. PP1 (1969), pp. 47–53.

BIBLIOGRAPHY

Collingwood, R. G., *An Essay on Metaphysics*, Oxford University Press, London, 1940.
Frank, Phillipp, *Philosophy of Science*, Prentice-Hall, Inc., Englewood Cliffs, N. J., 1957.
Hall, Arthur D., *A Methodology for Systems Engineering*, Van Nostrand, Princeton, N.J., 1962.
Machol, R. E. (ed.), *Systems Engineering Handbook*, McGraw-Hill Book Company, New York, 1965.
Searles, Herbert L., *Logic and the Scientific Method*, The Ronald Press Company, New York, 1956.

E. JOBST

SPECIFIC FEATURES OF TECHNOLOGY IN ITS INTERRELATION WITH NATURAL SCIENCE*

To determine the place of technology and of the academic engineer in the process of the scientific and technological revolution it is especially important to examine the connections between technology and natural science. There is very clear reference in the documents of the 7th Party Conference to the great responsibility to the community borne by the Socialist scientist and engineer for the planned, and thus deliberate, shaping of the developed system constituted by Socialism. For the academic engineer, *one* prerequisite for living up to this responsibility is an understanding of the theoretical problems of the interrelation between technology and natural science, because this is what determines, to a decisive extent, the capacity for applying the findings of modern science creatively for the benefit of our Socialist development.[1] The research and discussions conducted by Marxist philosophers attempting to define the nature of technology must therefore be aimed to a far greater degree than hitherto at investigating in detail the diversity of the connections existing in this sphere, since only in this way can the complexity of so significant a sphere as technology be grasped.

This article is based on J. Müller's work on defining concepts in technology[2], and its aim is to depict certain problems in the relation between technology and science which are a generalization of specific investigations in the field of thermodynamics. Hitherto, the connections between technology and science have often been regarded in a very one-sided way, as is most drastically apparent in the common description of technology as applied science. K. Tessmann rightly points out that despite their essential unity there do exist differences between scientific and technological research[3]. To draw attention to a few of these specific features of technology, which are the result, and yet at the same time the precondition, of its interrelation with science, is the purpose of this article.

The cooperation between technology and natural science must be conceived of as a dialectical interrelation, i.e., the interaction of findings in the two disciplines is a necessary condition for theoretical progress to

F. Rapp (ed.), Contributions to a Philosophy of Technology, 124–133.

be made by one or both of the sides involved in this process of mutual influence. The scientific and technological findings involved in the process play an *active* part; by reason of their relative autonomy they exercise a mutually stimulating influence on theoretical development and accomplish the process of mutual permeation by independent transformation, modification and specification of the theoretical notions operating in each particular case. To point out that in the process of the scientific and technological revolution science and technology are becoming more and more closely entwined is only of value if accompanied by a demand for investigation of the factors at work in the two scientific disciplines which make this process possible and condition it. The tendency of the two disciplines towards growing unity does not mean that they are coalescing but that within their interrelation they are affecting each other in a more comprehensive way. A prerequisite for planned, deliberate execution of the scientific and technological revolution is thus a precise definition of the specific theoretical features of technology and science. Naturally, none of the distinctions is ever absolute in the sense of rigid separation and delimitation. So when, with regard to *technology* in particular, certain features in this interrelation are stressed here as peculiar, they are ones which are predominantly, but not exclusively, typical of technology.

First, it should be pointed out that the interrelation between technology and science passes through various historically determined stages of development and is an expression of the level attained by the two disciplines, the mutual influence, and thus the integrating permeation, of the two spheres becoming the more intensive as the specific features of scientific and technological research crystallize more clearly. Technology, which came into being relatively late in comparison with the classical natural sciences, nonetheless developed right from the start in an active interrelation with natural science. Without doubt, as K. Tessmann[4] notes, natural science constitutes the foundation for technology; but it must be added that technology develops in an active relationship with science. One manifestation of this is the fact that the gradually emerging beginnings of technology are not only the result, but also a precondition of a creative technical application of scientific findings, in that they reveal the theoretical content of the technical problems and at the same time indicate the necessity for natural science to permeate them. Without the often laborious attempts of such men as Watt, Coriolis, Poncelet, Carnot, Redtenbacher, Weisbach, Reuleaux and others to elevate pursuit of technical aims to the plane of theoretical reflection, it would not have been possible to bridge the gap between natural science and technical requirements. The scientific knowledge available could not exercise direct influence on the development of technical procedures, for this the specific science of technology had to emerge as an intermediary converting the findings of the natural sciences into technological theory and making them available for the solution of *technical* problems, or setting the natural sciences tasks to solve *theoretically*. It must be stressed in this connection that the technological sciences only became definitely established as independent disciplines in the "conditions produced by the Industrial Revolution"[5], but at the end of the 18th and the beginning of the 19th centuries there are already investigations in the fields of mechanics, the theory of the strength of materials and the theory of heat which indicate that machines make theoretical reflection a necessity, that empirical comparison, combining and trying out possible solutions on technical objects are procedures which become more and more inadequate for the achievement of technical improvement. This does not yet constitute a systematic foundation for technology, but the theoretical and methodological preconditions emerge.

A second feature of the interrelation between technology and natural science in the emergence of technology is the fact that technical developments and the incipient theoretical reflection on them in technology directly stimulate the development of natural science. A striking example of this is the emergence of thermodynamics[6]. The triumphal progress of the steam engine in the first decades of the 19th century, its enormous economic influence, imparted urgency to the demand for greater efficiency. Precise evaluation of steam engines, particularly as regards the technique of heat utilization, became a more and more pressing need. Watt's work on the steam engine had already suggested that progress must lie particularly in the direction of thermodynamic research, but it was the French engineer Sadi Carnot, in particular, who perceived quite clearly that further improvement of heat engines depended primarily on theoretical understanding of the heat processes in the engine. He it was, too, who in 1824 with his reflections on the ideal cycle instigated a new stage of theoretical generalizations in the field of heat engines, and thereby directly created the necessary conditions for a theoretical mastery and scientific treatment of the thermodynamic problems of the steam engine. With the discovery of the first law of thermodynamics by a number of researchers a start was made, proceeding from various positions and on different theoretical levels, on the task of a general scientific investigation of heat processes, which Carnot's work had shown to be an essential object of theoretical endeavour. Here the emergence of a new scientific discipline was already apparent; the thermodynamic research became dissociated from direct application to specific technical problems. When Clausius and Thomson laid the systematic foundation of thermodynamics around the middle of the 19th century, the first complete scientific theory of thermodynamics came into being. In the constant interaction of scientific findings and the results of theoretical investigations on a technical basis, there thus developed a system of thermodynamics that constituted a component of natural science in general, but also furnished in turn the basis for more profound theoretical investigation of the *technical* aspect of heat processes. In the second half of the 19th century there developed an increasingly independent *technological* theory of thermodynamics. Engineers played a particularly important part in its development, Rankine, Hirn, Zeuner and Linde must suffice here as examples.

A glance at the history of the emergence of technology thus confirms that right from the beginning it had to work independently at solving problems determined by its specific field of application, that it developed and became established not only on the basis of natural science but in a dialectical interrelation with it and that this process of interaction was marked by various phases of original integration and differentiation.

It is already clear from this short historical consideration of the interrelation that technology actively affects natural science. The following specific features of technological theory are intended to draw attention to at least a few aspects of research and development work in technology which are both a precondition and a result of its dialectical connections with natural science.

(1) Technology has the task of incorporating in theoretical work the parameters of the functions of technical constructions. Its theoretical investigations must take into account right from the start the question of technical viability, i.e. they must regard ways and possibilities of technological adaptation of the theory as direct components of the research work. The technologist has to express technical problems and require-

ments in a form sufficiently general for them to be capable of theoretical investigation. It goes without saying that this requires extensive knowledge and creative application of this knowledge. Thus, to give just one example, many economically important problems concerning the exchange of heat and matter in technical processes are at present only familiar as difficulties which have not yet been mastered technically, and extensive research has still to be done in order to elevate them to the plane of theoretical investigation. Only on this plane of a theoretical mastery of technical problems does it become possible to select general scientific findings in accordance with technical requirements and adapt them in terms of technological theory. The derivation of the concept of energy by means of technological thermodynamics makes it clear that the physical import of the first and second laws of thermodynamics, although a precondition for technological research on problems of energy conversion, can never replace the creative systematic work of the engineer, who has to work out conclusions with a sound theoretical basis which have to make the concrete diversity of the technically conditioned processes of energy conversion capable of numerical calculation.

In the process of adapting general scientific knowledge in terms of technological theory, technical tasks are solved with the aid of scientific theory. At the same time, however, the research worker in technology often encounters technical problems which cannot yet be solved completely because the corresponding scientific research has not been done. It is clear that in this way technology exercises on natural science a stimulating influence which is an essential factor in the planning of scientific perspectives and basic research in accordance with the aims involved in the establishment of a comprehensive Socialist system.

(2) Due to the fact that the results of technological research have to serve the purpose of evaluating existing technical constructions more precisely and improving them, or of constituting a foundation for new technical advance, problems constantly crop up in technological research that involve new questions, ones to which natural science cannot provide answers. Thus the technologist is constantly faced with the double task of seeking to discover, in relation to the requirements posed by specific technical problems, which scientific findings offer scope for technical adaptation, and of taking scientific results as a basis upon which to formulate new technological principles and examine the possibilities of

applying them in practice. An additional factor is the difficulty that not every solution that is possible in theory is acceptable to the technologist. The transformation of scientific knowledge into technological theory thus involves as a rule making a selection of findings from various scientific fields and combining them. The academic engineer must decide which item of scientific knowledge helps to provide the best solution to his technical problem, and he must find out what combination of such knowledge can contribute to theoretical mastery of the technical problem. This decision must not rest solely on scientific considerations. Technological theory and the actual stage of development technology has attained exert decisive influence on the feasibility of technological application of scientific knowledge. Moreover, and this is no less important, an optimum solution can only be achieved if technological theory takes account of social conditions. Thus with each task of new technical development or improvement of existing facilities, such factors as energy supply, economic, commercial and social aspects, and considerations of work psychology, among others, must be incorporated in the theoretical technological reflections on which decisions on the use of scientific findings are to be based. It presumably goes without saying that in our Socialist system the engineer consequently needs not only a thorough education in natural science but extensive sociological knowledge as well.

It is thus possible to say that the problems of technological theory are largely of a specific nature and do not lie within the sphere of the tasks of natural science. The progress of technology does not eliminate these specific kinds of problem but is constantly reproducing them on a higher level. The ever-increasing mutual permeation of technology and science will not lead to problems of technological theory being solved by science. The trend – already apparent today – is rather for science to amass, with increasing reference to specific aims, a store of theory relevant to technological research, while technology is increasingly in need of scientific research in order to be able to find theoretical solutions to its own problems. On the other hand, technology is tending more and more to set science theoretically formulated tasks. This means that it is a basic condition for the deepening of the interrelations between technology and science that the latter should always be some distance ahead. It thus becomes apparent that the mutual permeation of the two disciplines does not lead to disappearance of the differences between the kinds of problem they face. On the contrary, only if the two distinguishable spheres of activity are clearly delimited, can the reciprocal relations be deepened and rationally planned cooperation take place between technologists and scientists.

(3) In technology there is an indivisible unity of theoretical research and development work. Technical practicability is, right from the start, an integral aspect of the theoretical deductions in technology, which contain a multitude of parameters determined by the functions of the technical constructions. Technological theory is therefore constantly influenced by future development work, as the object to be developed, i.e. realized in technical form, is anticipated conceptually. The materialization of technological theory, converting it into a real technical construction, is the concern of technological development work. Just as account is taken beforehand of technological development in basic and general technological theory, concrete application of the theory is itself a problem that has to be dealt with *theoretically*. Without characterizing precisely the individual steps and stages of theoretical work in the process of technological development – there is still a lack of detailed investigations on the subject – it is nonetheless possible to point out in principle that technological development constitutes the necessary complement of technological theory in the process of transformation from the abstract to the concrete. This process is constantly throwing up both new theoretical questions, arising out of the increasingly specialized attention paid to concrete technical conditions, and knowledge which is derived from the actual function of the technical constructions developed and so reacts on the general theory on which the development was based, modifying, complementing and perhaps correcting it. Thus today, for instance, basic theoretical statements on the effectiveness of particular cooling installations in cooling towers are only possible if, on the basis of a tentative theory, the manifold and often very complicated factors of influence are traced and analysed during the process of technological development, and thus comprehended in theoretical terms.

One characteristic, among others, of the present stage in the development of technology is that technological theory and development are merging more and more. This is apparent in the fact that future technological development as determined by social needs, is receiving increasingly thorough attention at the theoretical stage, that the theoretical preparation and carrying out of the development work is becoming better and better, that consequently the development work is coming to exercise more and more influence on the perfecting of the theory and, finally, that technological theory is a product of the dynamic unity of findings of general research and of development research.

(4) One important specific task of technology consists in preparing its theoretical findings with a view to rational utilization. It is an inherent feature of technology that its activity is not confined to research institutions but extends to a multitude of levels in the direct process of production. But this direct daily involvement would be impossible if in order to cope with a problem occurring in the course of production it were necessary every time to call upon the whole complicated system of theories. For this reason, one concern of technological research must be to present its findings in such a way as to facilitate swift, variable theoretical treatment of technical problems, in accordance with changing circumstances. This means that the task is not straightforward simplification of the theory but the much more difficult matter of working out (relatively) simple relationships embodying the essential theoretical content.

This concern of technology is especially apparent in diagrammatic presentations. For instance, to evaluate the changes of state of steam and moist air purely by means of calculation leads to extremely complicated equations, the solution of which is so time-consuming as to be no longer economically feasible for the engineer. In diagrams, on the other hand – and therein lies their significance – the heat energy processes in these changes can be traced with the same accuracy, but much more easily. In this way, theoretical knowledge which is difficult to grasp purely by means of calculation is, as it were, made manageable again. All the essential parameters characterizing the patterns of the processes to be investigated can be depicted. Thus, for instance, the i, x-diagram expresses the thermodynamic properties of air-steam mixtures consequently and precisely. The effect of altering one component can be traced directly and results can thus be attained swiftly. In addition, one diagram can cover a large number of different possible combinations. A further important factor is the illustrative quality of diagrammatic representations. True, modern thermodynamic diagrams are not illustrative in the sense that indicator diagrams are, what they offer is more an illustration of ideas. To anyone ignorant of the physical and technical features of the heat processes involved, the diagram does not mean anything. It is, in a certain respect, a model, which constitutes an illustration of the theoretical, analytical investigation of a physico-technical process, but on a level which does not consist in copying the phenomena of the real process but in illustrating graphically the conceptual analysis of this process in its fundamental structure and dynamics, which are not perceptible to the senses. In this sense diagrammatic representation is not a purely formal system that can be interpreted in any way at all, but preserves the meaning of the relations by registering the structure and dynamics of quite *specific* processes. In this way the diagram corresponds to the engineer's way of thought. For one thing it permits theoretical formulations of basic relations in physico-technical processes to be handled directly and illustrated, as described above, it being possible to trace the course of the process in the diagram and so acquire new knowledge without setting the whole extensive machinery of mathematical calculation in motion. Secondly, in the diagrammatic representation general theoretical knowledge concerning the physical and technical changes is combined with knowledge of the conditions under which the heat engineering process takes place in a particular case. It is this blending of investigations in the fields of general science, technological theory and concrete technical application which is the basis of the great significance of diagrams in technological research and practical application. Modern diagrams of applied thermodynamics are a convincing proof that the creative research activity of technologists is an indispensable precondition for the interrelations of natural science and technology.

This list, which is by no means complete, of the specific features of technology can of course only constitute an approach to a closer definition of the interrelation of technology and science. More precise conclusions can only be attained through analysis of the manifold interconnection of scientific and technological knowledge with the aid of concrete studies. It would seem particularly necessary to conduct investigations which clarify the interaction of natural science and technology in the process of forming theories and hypotheses in both disciplines. Closely connected with this is the question of the relations between the laws of natural science and those of technology. J. Müller's[7] observations on these problems suggest that a precise investigation of the process of theory formation – especially in technology – is an essential prerequisite for further examination of the specific features of technological laws as an expression of the interrelations of nature, technology and society. And for the now urgently necessary research on the integration of technological and sociological knowledge, too, a thorough examination of the formation and development of technological theories is indispensable.

The efforts of Marxist philosophers towards a profounder grasp and clearer differentation of the specific features of technology and the dialectical interrelations between this and other fields of knowledge are thus an important contribution towards making society more fully conscious of the tasks of technology in an advanced stage of Socialist development.

Karl-Marx-Stadt

NOTES

* Originally published under the title 'Spezifische Merkmale der technischen Wissenschaft in ihrem Wechselverhältnis zur Naturwissenschaft', in *Deutsche Zeitschrift für Philosophie* **16** (1968), 928–935.

[1] Cf. W. Ulbricht, *Die gesellschaftliche Entwicklung in der Deutschen Demokratischen Republik bis zur Vollendung des Sozialismus*, Berlin 1967, section VI.

[2] Cf. J. Müller, 'Zur Bestimmung der Begriffe 'Technik' und 'technisches Gesetz' ', *Deutsche Zeitschrift für Philosophie* **15** (1967), 1431–1449.

[3] Cf. K. Teßmann, 'Zur Bestimmung der Technik als gesellschaftliche Erscheinung', *Deutsche Zeitschrift für Philosophie* **15** (1967), 509–527.

[4] Cf. *ibid.*, p. 524.

[5] Cf. *ibid.*

[6] Cf. E. Jobst, 'Philosophische Probleme des Wechselverhältnisses von technischer Wissenschaft und Naturwissenschaft', *Wissenschaftliche Zeitschrift der Technischen Hochschule Karl-Marx-Stadt* **9** (1967), 81–92.

[7] Cf. J. Müller, *op. cit.*

D. TEICHMANN

ON THE CLASSIFICATION OF THE TECHNOLOGICAL SCIENCES*

My concern, actual classification of the present-day technological sciences, has not yet been touched upon. I do not want to speak about what general principles have to be considered in the classification of scientific disciplines but about the extent to which the technological sciences have already been classified – technologists have so far worked on the problem without the aid of philosophers – and about which aspects of approaching the task of classification should be taken into account in the specific case of the technological sciences.

Prof. Harig has shown very clearly in his lecture how the historian has arrived at a concept of the mode of classification practised in the past by examining the teaching structure of the times.

Our problem is concerned with the same connection, but in the other direction. We are confronted with the demand for a classification of the technological sciences *in order to impart greater precision to the teaching structure.* There are two general aspects of classification of the technological sciences. One is the classification of technology as a whole as distinct from the natural and social sciences, and the other is the – to use Herlitzius' term – 'internal' classification of the technological sciences themselves.

I should like to make two remarks in connection with the first aspect:

I too regard the object of a scientific discipline as consisting in the specific laws applying in the field involved, it being the task of the discipline concerned to investigate and record them, and not, as stated in the theses[1], in the totality of the relations and phenomena within one field. Otherwise everyday knowledge and science would have the same object. Also, it would in my opinion be very difficult even without this restriction to distinguish between natural and technological sciences, since they may examine the same objects; however, the relations they examine involving these objects are different – and it is precisely these specifically distinct laws governing the same objects that justify the autonomy of various disciplines.

F. Rapp (ed.), Contributions to a Philosophy of Technology, 134–139. All Rights Reserved

Furthermore, I am of the opinion that the principle of 'coordination and subordination', applied to technology, must necessarily lead to the latter being subordinated to the natural and social sciences. In the theses natural sciences, social sciences, technological sciences and other sciences (somewhat obscurely: 'Sciences such as, e.g. ...') are adduced in that order. But if one were aiming to coordinate in this manner, and if one proceeds from the assumption that classification must be derived from an objective structure of reality, then technical phenomena would have to be regarded as an autonomous field on *a par* with natural and social phenomena. This does not seem possible to me.

Without understanding by subordination a judgement on the scientific level or social usefulness, I am of the opinion that technology is subordinate to the natural and social sciences. This does not mean that technology is only applied natural science or anything like that. But it is based on natural science; today still predominantly on physics.

What follows from the general aspects that is relevant for the internal classification of the technological sciences? How could such a classification be undertaken?

The first possibility is a *historical classification.* If one takes the historical qualities of technical development as a basis for distinguishing technological sciences, then tool, machine and automation engineering would be one's initial categories. Since the technological sciences only came into being with machinery – later still, really! – this historical division does not lead us far as a basis for a classification of the technological sciences. One would need to establish a historical division within the general concept of machinery. Max Bense attempts something like this. He speaks of the 'classical machine', that which performs work and generates energy. This would produce a complex of sciences which would be concerned with the principles of the lever and with the generation and transmission of steam power, or something of that kind.

His second stage is information and communication technology, the characteristic elements of which would be the punched-card system, radio and television.

Which sciences would deal with this stage?

In his third stage the main element is cybernetic technology, characterized by sciences concerned with control and storage.

A second possibility would be division of technology according to the *natural laws* essentially embodied in it, i.e. roughly into physical, chemical and biological technology. This would produce a division of the scientific disciplines into technical physics, technical chemistry etc. Since at present biological technology can be regarded more or less only as a visionary prospect, and the main emphasis is on physical technology, this latter would now have to be subdivided into mechanical, heat, electrical, optical engineering, atomic or nuclear engineering, and the scientific disciplines, too, would have to be divided accordingly.

This would perhaps be a 'logical' classification of technological sciences – a risky concept, which I may be permitted to use in this way here.

The whole idea, however, presupposes a classification of technology according to natural laws. Thus one could speak straightaway of a classification of the technological sciences in the form of a copy of a classification of the natural sciences.

A third possibility would be division according to *branches of production.* This would produce (1) agricultural, forestry and fishing technology; (2) mining and metallurgical technology; (3) civil and construction engineering and the production of building materials; (4) transport technology; (5) power-generating technology; (6) machine construction and electrical machine construction technology; (7) light industry and food industry technology; (8) administration technology and (9) equipment construction technology.

If one adopts this approach, one obtains a division not of the technological sciences but only of technological teaching subjects. It seems, in my opinion, clear that the present system of division in colleges of technology, faculties and institutes has been built up predominantly according to this principle.

This kind of pragmatical division of the technological sciences in accordance with the form in which they are taught is very limited, since training in all the engineering subjects is tending to be grounded to an increasing degree on the same fundamental knowledge.

A fourth possibility would be to divide technology according to its function within a branch or process of production.

I would remind the reader that Marx makes a division of this kind when he distinguishes within the general concept of machinery

(a) the machine tool and work machine,

(b) the transmission mechanism and
(c) the locomotive machine
(see *Capital*, vol. 1, p. 389).

This division of Marx's has been superseded in the course of technical development, because today nearly every individual machine fulfils all three of these functions itself (today, for instance, every machine tool has separate motor drive; the separate transmission mechanism of the time before discovery of the electric motor has long since disappeared). But a similar division of machinery takes place again if the complex technical constructions are combined according to their basic patterns.

In this sense it is customary today, too, to make a rough division into the technology of the transformation of materials, that of the transformation of energy, and possibly the technology of information- and data-processing. This would produce the following division of technology into

(a) engineering mechanics and related disciplines (which will not be considered in any more detail here)

(b) electrophysics – a venturesome concept – and

(c) the technological sciences concerned with regulation and control processes.

Our fourth possibility, considered above, also brings us to a pragmatical classification of the technological sciences. The result, however, evinces a certain similarity to the other possibilities, in particular those considered first and second.

I should further like to consider a fifth aspect, that of the construction principle of the machine itself. This is, of course, closely connected with the function of the machine in the process of production. For this reason I should like to regard this fifth aspect only as a refinement of the fourth.

For our purposes a very much simplified scheme, which could be extended at will, could be drawn up, distinguishing between

(a) machinery for the extraction of raw materials, conveying machinery, incineration, heating, coaling and lighting installations;

(b) energy machines (wind- and water-power engines, ones using the energy of the sun, steam and internal combustion engines, generators, electric motors and reactors);

(c) processing machines for coal, ore, stone, soils, metal, wood, plastic, paper, leather, textiles, food and so on;

(d) measuring equipment, data-processing machines.

Does this scheme lead to a classification of the technological sciences? I think not.

For one thing, it goes too far, for there is no technological science specializing solely in the manufacture of textile machines.

Furthermore, technologists themselves say that the type of construction cannot be differentiated to the same degree as the application of the technical principles involved. A cement-mixer and a machine for kneading dough have, as far as I know, a similar structure, but I have never known the two of them assigned to the same category of technical appliances.

It is apparent from this problem that classification of technical constructions can certainly constitute a starting-point for classification of the technological sciences, but that the two are not identical or congruent.

What conclusions can be drawn from all our considerations so far?

(1) There are certainly various possible ways of classifying the technological sciences. The pragmatical aspect will emerge whichever possibility one adopts, regardless of whether the technological sciences are being distinguished for teaching purposes (What is the engineer being trained for? What technical field will he be concerned with? What technical appliances and processes must he understand and develop? The result is the major subjects and their syllabuses.) or whether their classification is being derived from that of technical constructions (in which case it has to be taken into account that the technical constructions are created by man, i.e. classification of them is preceded by a structure of human aims and purposes). The only meaningful classification of the technological sciences will be one which takes into account both the objective structure of technology, and the classification of the natural sciences and the teaching structures.

(2) The pragmatical aspect of classification of the technological sciences prevents unlimited ramification of the classes defined. Despite all the specialization in research on functional relationships in technology, both production and teaching are aimed at creating, respectively, a unified complex machine, and the unified training necessary to build and work it. For this reason the structures both of teaching and production lead not so much to individual, precisely delimitable technological sciences as to complexes of technological science (mechanical, electrical, civil and construction engineering etc. as faculties are not individual sciences but scientific complexes).

(3) If one considers the increasing complexity of modern technical installations and machines, the adaptability that has come to be a necessary feature of the engineer's training, and the larger part being played by basic knowledge of the natural sciences in an engineer's training and activity, and seeks their common significance with regard to classification of the technological sciences, then it becomes necessary, in my opinion, to distinguish particularly between *basic technological disciplines* and special fields of technology. This produces two levels of classification.

With mathematics, physics and chemistry as the natural sciences constituting the foundation of the engineer's training, the basic technological disciplines should comprise, as far as I know, theoretical mechanics, theory of the strength of materials and mechanical engineering; thermodynamics, hydrodynamics, general electrical engineering; theory of gears, theory of projection, kinematics, theory of machine parts, and theory of design as a whole (in as far as it has been elaborated as a recognized discipline). Here natural laws and their functional application are the predominant criteria of classification.

The special fields of technology are then built up on these bases, the specific nature of the technical constructions involved being the decisive factor in delimiting these fields. Here the basic disciplines are applied for particular practical purposes. For this reason I refrain here from speaking of specialized disciplines.

Dresden

NOTES

* Originally published under the title 'Zur Klassifikation der technischen Wissenschaften', in *Klassifizierung und Gegenstandsbestimmung der Geowissenschaften und der technischen Wissenschaften* (Freiberger Forschungshefte D 53), Leipzig 1967, pp. 199–203.

1 [The theses referred to are those which served as a basis for discussion at the colloquium for which this paper was written. They appear not to have been published.]

T. KOTARBIŃSKI

INSTRUMENTALIZATION OF ACTIONS*

The wealth and rapidity of improvements in apparatus is a characteristic of human technology and constitutes, together with language, the principal and externally observable functional trait of the species *Homo sapiens*. It is to those distinctive functional characteristics that that species owes its successes in mastering the forces of Nature on our globe. What achievements in making actions more effective are due to persistent instrumentalization, that is the increasing share of apparatus and the growing importance of apparatus in agenthood processes? It is my intention to approach these issues now in the belief that, indirectly, I shall thereby review the principal recommendations concerning the improvement of actions through instrumentalization.

Now, first of all, owing to instrumentalization we obtain products and perform works which would be impossible without instruments or adequate improvements of such. As for products, we encounter such artefacts everywhere: timber processed for making furniture, concrete slabs for making pavements, textiles for making clothes, etc. They are common products met with in everyday life. And what about results? There is no lack of them, either; let us mention here only the use of electric power for lighting and driving trams and trains. Now, there are examples showing how far instrumentalization enables us to extend the field of human actions, especially in those spheres where the results obtained are measurable. Levers and pulleys enable us to transport solids which are enormous when compared with those solids which we are able to move directly by the strength of our muscles. Modern devices can launch missiles across the oceans (compare with that the records of the best discus or javelin throwers). Aircrafts carry loads of passengers and goods over stupendous distances in a stupendously short time, as compared with the quickest runners. Telegraph and radio convey information throughout the globe with a velocity that comes close to the maximum velocity existing in Nature, whereas without apparatus we can send signals at the most within the reach of human sight or human hearing.

Indeed, man has succeeded in perceiving visual stimuli coming from celestial bodies that are millions of light-years away. Thus, as regards reception, man is in touch with the universe on a cosmic scale of distances. True, this is not accompanied by such a level of instrumental technique as would enable us to send signals to Arcturus, Vega or Betelgeuse, but nevertheless instrumentalization increases many times the perceptive abilities of the human eye. I do not mean here especially telescopes, which enable us to distinguish details of light pictures of distant bodies, although this, too, is a marvellous achievement of instrumentalization. I am here principally concerned with photography combined with telescopy: instruments, films sensitive to light, lenses and mechanical devices, reach incomparably further into the cosmic space than does even the most sensitive human retina. Magnifying glasses, apart from the function of bringing remote objects near, also perform – this time in microscopes – the function of making extremely small objects visible. Further, an X-ray apparatus reveals a foreign body which has penetrated a human organism. Thus the various types of apparatus make accessible to human perception certain objects which without their help could not be perceived by human senses.

This is so because certain instruments react to such small differences as are not distinguished by human senses. Hence, working by means of instruments ensures a much greater precision than working without them. Our intuitive evaluations of distances, weights, periods of time, etc., are extremely rough in comparison with the precise measurements made by means of triangulation instruments, precision scales, chronometers, etc.

With the aid of certain instruments, one works incomparably more economically than without them. There are instruments which amplify our impulses when transferring them from the object upon which our impulse has worked on to the appropriate material. The ordinary lever causes a force applied to the end of an arm which is *n* times longer to give an effect which is *n* times greater at the end of a shorter arm. Other instruments economize actions still more, since they serve to release large amounts of energy, as for instance when a slight pushing of the button explodes a dynamite charge and tears a hill apart.

The various instruments and technical installations in general enable us considerably to minimize interventions. I once had an opportunity to watch at a railway station on the frontier between two countries how pedestrians were protected against manoeuvering trains. In one country, a guard was posted to prevent people from crossing the tracks at moments of danger. In the other country, a tunnel was constructed under the tracks, so that the pedestrians were not exposed to the slightest danger of falling foul of a train, and no one had to watch the crossing. This is a telling example of a minimization of intervention, in this case through an engineering solution of a traffic problem.

The use of apparatus makes it possible to do many things 'at a single stroke'. A circulated printed text or a text broadcast over the radio conveys information to large numbers of people all of whom would have had to be informed individually if such an item of information had to be transmitted by word of mouth. The printing machine multiplies copies once the text has been set by the compositor. Moreover, the precision with which the shape of each letter is reproduced makes a printed text much more legible than script. Ships, trains and other modern means of transport convey from place to place much greater quantities of goods than an individual could carry. An appropriate installation makes it possible to light street lamps over the whole area of a city.

When we rely on effective instruments, we often obtain a much greater reliability and precision of action than when we manipulate directly by hand. By means of compasses, we can draw a much more exact circle than the hand can do unassisted by any instrument. Instruments, Francis Bacon long ago observed, to some extent compare with natural talents. Further, if an individual wants to be woken up early, to set an alarm clock is more certain than to rely on the hotel staff. A photograph gives a much greater probability of faithfully reproducing a likeness than does a portrait drawn or painted by hand.

It seems paradoxical to say that instruments can be more operative than one's own organs such as hand, jaws, tongue, etc. And yet let us compare the situation of a singer with that of a virtuoso. Consider the extent to which a good performance by the former depends even on an ordinary cold, which affects his vocal chords. And further, for a singer to pass a certain age limit may prove simply catastrophic. On the other hand, a musical instrument is not so whimsical and does not grow old so quickly. To put it in a more general way, instrumentalization – seen as the production and making use of apparatus of all kinds – gives our actions considerable independence of factors which otherwise would impede such actions or even make them impossible. For instance, the fabrication of buildings, heating installations and clothes (i.e. providing ourselves with certain apparatus) makes us independent of climatic fluctuations. Artificial lighting makes us independent of inconveniences and obstacles caused by darkness at night.

Finally, there is still another advantage in the instrumentalization of actions, which is of extraordinary importance for the co-ordination of actions by a number of agents. I mean here that instruments enable us to produce very many products which are so similar that one can very well replace another. This also enables us to build preparatory installations with great certainty as to their future smooth functioning in view of the standardized form of the various parts.

All these advantages of instrumentalization are enormous and universally known. Consequently, it is desirable to instrumentalize our actions on an increasing scale. Operativeness of instruments and of all apparatus must be one of the principal concerns of designers and constructors. Operativeness can be understood in two ways: subjective and objective. Subjectively, the more easy an object is to handle and to use for various actions, the more operative it is. Especially in this century, engineers and business managers have introduced a great many improvements with a view to increasing the operativeness of tools, etc. I refer now to such things as proper location of handles and levers so that they may be reached and moved freely, simplification of manipulations necessary for putting a machine into operation and controlling its work, and the like. A glaring example of differences into subjective operativeness is offered by a muzzle-loader and a breech-loader. And a pedal-operated sewing machine is also much more operative than a manually-operated one.

It would be difficult to overestimate the importance of an appropriate ordering of the component parts of composite installations (e.g. an installation which is a somewhat loose collection of elements), of the knowledge of such ordering, and of access to such elements. These factors signally increase the subjective operativeness of the installation, and neglect of them is probably the principal source of disorder and the resulting defects in collective action, popularly known as 'being in a mess'. A very good example of order is provided by well-run libraries, where every book is kept in the place assigned to it, and reflected by a corresponding entry and number in the catalogue which is a permanent source of information concerning the order in which the books are arranged. And here is an authentic example of 'a mess'. A clerk had to find a certain document among the papers left by his predecessor in a heap on the desk. He tried to find the document by the date on which it was signed, but the papers were not arranged chronologically. He then tried to find the name of the person concerned, but they were not arranged alphabetically either. Since he did not know the order, if there was any, in which the papers were arranged, he could not find the document, and the person concerned had to leave the office empty-handed.

Standardization of instruments and products is a measure similar to ordering. For instance, the fact that the threads of electric bulbs are standardized, as also are the fittings into which the bulbs are inserted, ensures an easy replacement of one bulb by another in any fitting. This enables serial production of both bulbs and fittings, the output of which may be correspondingly improved. The concept of standardization is also applicable, and has analogous practical values, when a group of men is treated as a set of instruments. This explains, for example, the introduction of curricula which are common to a given type of school. But in handling groups of people, standardization may result in antinomies, if it is forgotten that a collaborator is after all not an instrument. Standardization with respect to men is what the Germans call *Gleichschaltung*, and the universal aversion which that term evokes proves tellingly that collaborators usually resent any abuse of standardization tendencies with respect to them. (For details see the Chapter on co-operation.)

On the other hand, by operativeness in the objective sense of the term we mean, for instance, that the given apparatus has as few inherent defects as possible, and that consequently it requires a minimum of repairing, controlling, regulating, etc. Generally, the more operative a given installation in the objective sense of the word, the more efficient would be called an agent functioning as that installation functions (or, which amounts to the same thing, the less the functioning of that installation is blamed for inefficiency). A muzzle-loader is less operative than a breech-loader not only in the subjective sense, but also in the objective, since it takes much longer to load it after firing. The difference between the two resembles the difference between an unskilled or lazy worker and a skilled and vigorous one. Similarly, horse-drawn vehicles are much less operative as a means of rapid transport than motor vehicles. As regards the operativeness of instruments, installations, etc., technologists have to solve various problems, in particular those which refer to elimination and repair of defects, control, regulation, spare parts, etc. I shall now draw attention to two very general situations, in connection with the dependence of the design of such instruments and installations on the dynamics of formation of composite objects. First, an increase in the degree of organic unity in compound objects results in a tendency towards centralization. That tendency leads to the emergence within a composite object of a part of it on which all other parts depend unilaterally. The engine in a motor-car, the generator in a power plant are good illustrations here. Progress in that field takes place at the expense of the danger of a general paralysis of those parts which are thus unilaterally dependent. Such a paralysis occurs, for instance, whenever there is a defect in a power plant providing a town with electricity. Consequently, in those cases it is imperative to have available alternative facilities if and when the central installation fails. The latter problem, when not solved satisfactorily, can be very acutely felt, as in the case of municipal central heating. It is simply disastrous when the municipal central heating plant fails in winter.

Secondly, let us recall here the current reference to the 'impishness of small objects'. Quite often the functioning of a given apparatus depends on the presence in a definite place of some minute component part which, precisely because of its small dimensions, can easily be lost or mislaid; if this happens, the whole installation stops working, at least until the missing element is restored to its proper place or until a substitute is found and properly inserted. In such cases, it is advisable to have replacements in store. More important still is to make the apparatus independent of such component parts. Here is a simple example. Pneumatic cushions are very convenient for travelling. There are two types. In one type, the aperture is closed with a tiny peg which easily slips out and thus can easily be lost. The cushion then becomes useless, it being very difficult to replace the peg with any home-made device. The other type is incomparably more practical, since the stopper is in the form of a screw and is fastened to the cushion with a chain.

Let it be added that the concept of operativeness and everything which has been said in that connection, can be referred not only to instruments, but to other entities as well, in particular to teams and institutions. This point will be dealt with in greater detail when we come to discuss collective action.

Some caution is necessary concerning instruments. First, there is the danger of becoming excessively dependent on instruments, which in exchange for the services they render require adequate attendance. Their operation must be watched and corrected, and the movements of the man who takes care of them must be adjusted to the movements of the machine he uses. Man must, on an increasing scale, serve the machine which was intended to serve him. He must see to it that the furnaces do not become extinguished, that the rails do not become displaced, that axles and couplings are properly lubricated, that a broken piston is immediately replaced by a good one. All this is necessary but exacting, and engineers endeavour to alleviate it by way of increasing automation of the apparatus. Automation improves them so that they may become as self-regulating as possible and maximally approach the type of robots which are endowed with an increasing ability to react to the various stimuli, including the stimuli originating in deviations from standard functioning and causing self-regulatory reaction. But however the engineers may make the machine independent of human interference, to minimize the interventions of *homo faber*, the machine will always require guidance on the part of human agents and, in view of the fantastic rate at which all kinds of apparatus develop and become more and more complicated, the rational behaviour of man will depend to an increasing degree on the functioning of machines.

And these machines demand changes in the dispositions of human agents and force them not only to new efforts of inventiveness, to increased initiative, to growing watchfulness during invigilation, even in the case of pure invigilation, but also make it imperative for human agents to develop new automatisms and manipulative improvements as well as new kinds of impulses. Compare the difference between the type of skill formerly required from a clerk, and the type of skill now required from a typist. Compare the difference in skill needed by a photographer as compared with a portrait painter. The same refers to manipulations performed by the crew of a steamer as compared with the crew of a sailing ship, and, *a fortiori*, with the crew of a paddle boat. Thus, new skills must be acquired under the pressure of new compulsory situations which are due to instrumentalization. Yet that progress entails regression in various fields. Long ago, Jean Jacques Rousseau stated in his *Discourse on the*

Origin of Inequality that "our industry deprives us of that force and agility which necessity obliges him (a savage man) to acquire. If he had had an axe, would he have been able with his naked arm to break so large a branch from a tree?"[1] Here are some examples of regression of certain skills: the safety razor, easier and safer to use, is causing the disappearance of the skill of handling the ordinary razor, a skill that formerly used to be very common. The X-ray diagnosis of pulmonary diseases is causing a decline in the art of auscultation. A study of old documents and private letters and the comparison of such with the ugliness of contemporary scribble reveals the decline of calligraphy. Handwriting is more and more being relegated to the rôle of draft-making, and very few people, even at school, bother to develop a fine, legible hand. A century ago periodicals used to be amply illustrated by xylography, but now photography has completely displaced wood-engraving in the matter of documentary value, cheapness, reproducibility, and so on. Consequently, xylography would have completely disappeared had its rôle been purely informational. If it survives, and even experiences a renaissance, that is because from being a transmitter of information it has been transformed into a vehicle of expression of ideas of the black-and-white artist. It seems highly desirable that the consequences of instrumentalization of actions should take, whenever possible, the form of transformations in the applications of skills developed before instrumentalization, rather than the form of abolition of such skills. Rather the fate of the skill of wood-engravers than that of the skill of calligraphers.

There is still another praxiological danger resulting not so much from instrumentalization as from the rapid progress of such, especially if instrumentalization is not accompanied by parallel adaptative changes. For instance, a new machine makes old installations obsolete. New machines force the old ones to the wall, and thus deprive trained people of their former employment. The fewer the numbers needed to operate the new machines, as compared with the old ones, the more violent is the process. This often leads to crises, under the free competition system, where employment is determined by the supply and demand of manpower. Planned economy can remedy these deleterious effects by purposefully distributing manpower and re-organizing co-operation. Efforts to slow down the progress of instrumentalization usually fail, and it is even counter-purposive to renounce such instrumentalization as has already been achieved. There can be no return to childhood; at the most, a process of infantilization may occur. And a society which dared to start to reject progress offered by instrumentalization would soon face the danger of attack by its neighbours. Here is an example, caught so to say *in flagranti delicto*, of counter-purposive action in the form of a deliberate renunciation of a previous instrumentalization. In a library, it was decided to dismantle the indicator, a device which by means of a simple manipulation showed whether a given book was in or had been lent out. The amount of time and effort saved by that device is obvious. The answer to press criticism of the anti-progressive character of that measure was to the effect that it had been dictated by a desire to maintain direct contact between the public and the library staff. If such an argument were to be allowed weight, we could eliminate thermometers and X-ray apparatus in hospitals, because these replace direct contact between the patient and the physician, who need not measure temperature by placing a hand on the patient's forehead and need not auscultate by putting his ear to the patient's back. Better learn to preserve direct contact with the reading public without unceremoniously destroying the achievements of predecessors.

Notice, now, an apparent paradox. An inhabitant of a somewhat backward country, which by intense imitative activity is trying to catch up, travels in a country known for important technical improvements. He is surprised to find there telephones, railways, etc., worse and more primitive than the comparable services in his own country, and may even be inclined to conclude that his own country is more advanced. Yet the paradox can be easily explained. In that advanced country, telephones, railways, etc., had, as a result of remarkable enterprise and creative spirit, been introduced earlier than elsewhere. The imitators had taken over these inventions, added some improvements and, working in a technological vacuum, introduced them from the outset in such an improved form. It might be expected that the first country would then immediately introduce in its own territory those improved forms; but they would have no technological vacuum to work in; on the contrary, they would have to take into consideration existing installations, functioning well enough, so that it would be uneconomical to replace them by the improved ones. As a result, a creative initiator, full of enterprise, has at certain stages less efficient apparatus at his disposal than his imitators have. Does it then pay to be creative? Is it not more expedient to wait for the results of other people's creative work and then simply to imitate them, at the most introducing certain additional improvements? In my opinion, enterprise and epoch-making creative work after all do pay in the long run. This is so because one has earlier at one's disposal installations and devices which are essentially new, and when these become somewhat obsolete as compared with those of the imitators, then again, after a transition period, creative enterprise will introduce such new essential improvements that the imitators will find it extremely difficult to keep abreast of the real inventors.

Warsaw

NOTE

* First published in *Praxiology – An Introduction to the Science of Efficient Action*, Oxford etc., Pergamon Press 1965 (original Polish title 'Traktat o dobrej rabocie'), Chapter IX (= pp. 125–132).

1 *A Discourse on the Origin of Inequality*, in J. J. Rousseau, *The Social Contract and Discourses*, London, 1938, Dent and Sons, 'Everyman's Library', 9th ed. (transl. by G. D. H. Cole), p. 178.

M. ASIMOV

A PHILOSOPHY OF ENGINEERING DESIGN*

I. DEFINITION OF ENGINEERING DESIGN

Engineering design is a purposeful activity directed toward the goal of fulfilling human needs, particularly those which can be met by the technological factors of our culture. The satisfaction of these needs is not peculiar to engineering design; it is common to much of human activity. Earning a living by serving the requirements of others is one of the chief characteristics of the modern social environment.

A designer does not usually produce the goods or services which immediately satisfy a consumer's needs. Rather, he produces the model which is used as a template for replicating the particular good or service as many times as is required. A design may be of a pattern on wall paper or of a garment in the world of fashion. If the producer believes that a sufficient number of customers will be satisfied by replicas, then production of the item or service may follow. In the course of production an error made by the producer in fabricating any one replica may lead to a rejection; but an error in design, repeated in all replicas, may lead to an economic misadventure of serious proportions. The designer's responsibility is therefore large.

As a profession, Engineering is largely concerned with design. What distinguishes the objects of engineering design from those of other design activities is the extent to which technological factors must contribute to their achievement. Every design activity that finally leads to a physical embodiment of the designer's conception must perforce make some use of technical factors. The key is the level of sophistication required in the manipulation and application of these factors and the extent to which a well-developed understanding of the underlying physical phenomena is necessary. Thus the sculptor, who produces his design in marble or bronze, is principally concerned with the aesthetic factors of form, shape, and texture. He is most likely aware of the technical behavior of his medium; but this is not his primary concern. It is sufficient for him to know as much about the technical factors which impinge on his design as would a highly skilled craftsman. To generalize, if a design can be accomplished properly with a simple technology, or with one that can be reduced to a routine that can be learned at the craftsman's level, then engineering design is not required. Engineering design *is* involved when the appropriate technology is complex and its application not obvious, and when the prediction and optimization of the outcome requires analytical procedures. Engineering design almost always requires a synthesis of technical, human, and economic factors; and it requires the consideration of social, political, and other factors whenever they are relevant.

II. DESIGN BY EVOLUTION

The feature that seems most characteristic of our times is the rapid pace of technological development. Scientific discoveries, multiplying in frequency, become available for technological exploitation. Society, which in the past had tended to abhor rapid change, has become receptive of and eager, and at times even impatient, for new feats of engineering design.

In the past, designs tended to evolve over long spans of time. Devices or technical systems changed gradually as time went on, each change making a small improvement on the preceding model. The leisurely pace of technological change reduced the risk of making major errors. The circumstances rarely demanded, and consequently seldom elicited, the utmost skill and analytical capabilities of the designer. This was 'design by evolution'; the technical risks were small, and the stakes were proportionally small.

It is true that competition has long presented a stern and relentless challenge, but the contest, occurring on the commercial plane, was primarily waged in the market place. The designer was shielded by the salesman. Today the range of competitive action has widened. The struggle appears not only in the market place, but in the design office and the development laboratory as well. The challenge has moved to the technological plane, where the gradual and unhurried improvement of a product is now less likely to meet the demands of competition. Present circumstances require bolder, faster improvements. Consequently, the technical risks which engineering designers must face today are very great, as are the stakes – so large that they transcend the bounds of private interest and sometimes involve the entire national economy.

III. DESIGN BY INNOVATION

More frequently now than ever in the past, products are designed *de novo*. Following a scientific discovery, a new body of technical knowledge develops rapidly, the proper use of which may dictate an almost complete break with past practice. A new design is projected, based on ideas hitherto untried. The outcomes are shrouded in the obscurity of the future and blurred by the complexity of the technology. The risk of technical errors is immense. Every skill which the designer or the design team can muster in analysis and synthesis is scarcely enough.

The analytical tools which derive from the engineering sciences relating to fluid flow, heat transfer, electrical phenomena, etc. are well developed and need no elaboration here; but besides dealing with these physical phenomena, the designer encounters a host of problems which are peculiar to the process of design. They arise from the need of developing, organizing and evaluating information almost always in the face of uncertainty, from the necessity of taking into account the complicated interactions of components, from the constant requirement to make predictions in terms of design criteria, and from the need to work always within the constraints of an economic framework. We will speak of the analytical techniques which cope with these problems as the general methods and tools of design as distinguished from the specific tools which arise from the engineering sciences. It will be the general methods and tools of design, not the specific tools, that will be our concern in this text.

IV. PHILOSOPHY AND DISCIPLINE

Philosophy may seem like a forbidding word to an engineer; but it simply means, in its literal sense, a love of wisdom. More specifically, a philosophy implies wisdom that has been organized to form a usable intellectual structure. It is a body of principles and general concepts which underly a given branch of learning, a major discipline, a religious system, or any other important human activity; and it includes the application of these principles in the domain of their relevance. Thus we can speak of a philosophy of history, or of business, or of Christianity. Philosophy has another meaning: it is a consistent and integrated personal attitude toward life, or reality, or certain phases of them. Thus, for example, it could be one's attitude towards his profession, especially if this attitude is expressed in beliefs or principles of conduct.

A philosophy does not spell out the detailed action one should take in specific instances; rather, it deals with underlying principles, concepts, and general methods which are relevant to whole classes of problems. The broad principles, the concepts, and the general methods of the philosophy lead to the development of theories, laws, and rules and to detailed methods of applying them. The latter is the discipline of design; it derives from philosophy, and it deals with recognizable categories of problems rather than with the general or the very specific.

Whereas a philosophy forms an intellectual superstructure or overall strategy which molds and guides the development of discipline, discipline provides an intermediate intellectual structure or strategy which molds and guides the attack on categories of problems. The practitioner, when dealing with an immediate and particular problem, must develop from his knowledge of discipline a specific attack or tactic which resolves that problem.

V. PHILOSOPHY OF ENGINEERING DESIGN

To develop a philosophy of engineering design we must seek out those principles and concepts that are of the greatest generality, consistent with usefulness, and that can lead to a discipline of design. We must formulate a method whereby the discipline of design is applied in the most general sense. We will need also to establish a third element to take care of the evaluative function by setting forth a general critique that provides a way of measuring the validity and value of results in specific application.

The principles and concepts on which we must rely stem from the collective experiences of mankind. They represent, in an empirical sense, what we believe to be true. As a rule, we cannot prove by means of logic that any particular principle is true; we can only support our belief in its truth on the empirical evidence of experience. Hence it is inevitable that the choice of principles, and their formulation, will be colored and biased by our personal experiences and idiosyncrasies. For this reason there cannot be *one* philosophy of engineering design, any more than there can be *one* philosophy of history, or *one* philosophy of science. In architecture, for example, there have been in the past, as there are at the present time, many schools of thought, each representing a different philosophy with its adherents.

Although the choice and formulation of the principles that underlie a philosophy are subject to the vagaries of the individuals who construct it, the principles themselves must nonetheless be bound by the rules of logic. This must be so if we wish to exert the power of logic in applying them to the situations which are subordinate to the philosophy. The principles must, therefore, form a consistent set, so that one does not contradict another. They must be capable of expansion by logical combination and extension to form a larger body of derived principles on which discipline may find a secure foundation.

A philosophy that does not lead to action is sterile; it becomes an exercise without consequence. It must embody, if it is to be useful, a methodology whereby the principles can be applied in a disciplined way. This operational aspect is the chief burden of discipline; it involves the procedural, logical, and quantitative techniques that occupy much of the remaining text.

The origin of the philosophy is empirical, but its test is pragmatic. The solutions to which it leads must be good in the sense that they are useful. But good is a relative term that needs a specific definition especially tailored for each particular situation; its value needs to be measured in a way that is peculiar to each situation. Therefore, the philosophy must include an evaluative scheme which guides and enables the formulation of specific criteria of goodness. This evaluative element is essentially a feedback mechanism which serves to indicate how well the principles have been applied in the particular instance and to reveal shortcomings so that an improved application of the principles can be made.

In the view which is set forth here, a philosophy of engineering design comprises three major parts, namely, a set of consistent principles and their logical derivatives, an operational discipline which leads to action, and finally a critical feedback apparatus which measures the advantages, detects the shortcomings, and illuminates the directions of improvement. These ideas, abbreviated and perhaps oversimplified, are diagrammed in Figure 1.

We shall conclude this chapter with a listing of the principles on which subsequent discussions will rest. Other concepts and ideas which can be informally derived from this set are advanced as needed. The word *design*, wherever it appears in the following, will mean engineering design. The list is not intended to be a rigid set of formal postulates.

(1) *Need.* Design must be a response to individual or social needs which can be satisfied by the technological factors of culture.

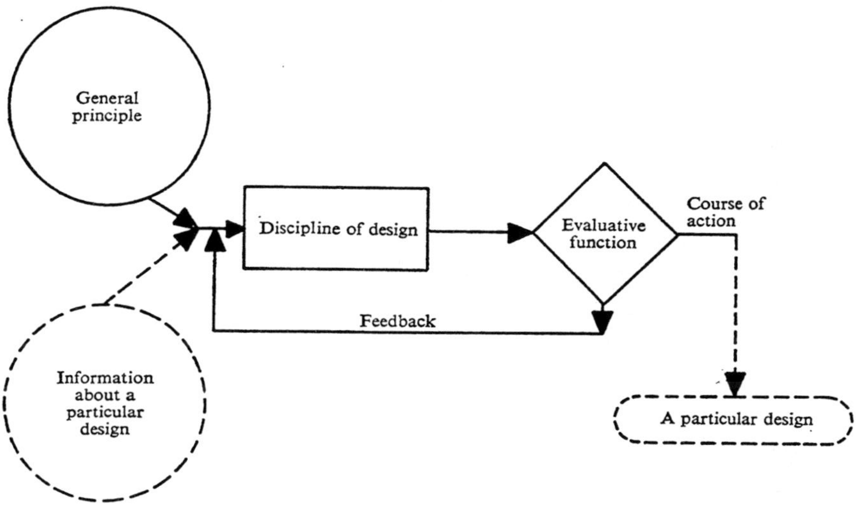

Fig. 1. Philosophy of Design. The feedback becomes operable when a solution is judged to be inadequate and requires improvement. The dotted elements represent a particular application.

(2) *Physical Realizability.* The object of a design is material good or service which must be physically realizable.

(3) *Economic Worthwhileness.* The good or service, described by a design, must have a utility to the consumer that equals or exceeds the sum of the proper costs of making it available to him.

(4) *Financial Feasibility.* The operations of designing, producing, and distributing the good must be financially supportable.

(5) *Optimality.* The choice of a design concept must be optimal among the available alternatives; the selection of a manifestation of the chosen design concept must be optimal among all permissible manifestations.

(6) *Design Criterion.* Optimality must be established relative to a design criterion which represents the designer's compromise among possibly conflicting value judgments that include those of the consumer, the producer, the distributor, and his own.

(7) *Morphology.* Design is a progression from the abstract to the concrete. (This gives a vertical structure to a design project.)

(8) *Design Process.* Design is an iterative problem-solving process. (This gives a horizontal structure to each design step.)

(9) *Subproblems.* In attending to the solution of a design problem, there is uncovered a substratum of subproblems; the solution of the original problem is dependent on the solution of the subproblem.

(10) *Reduction of Uncertainty.* Design is a processing of information that results in a transition from uncertainty about the success or failure of a design toward certainty.

(11) *Economic Worth of Evidence.* Information and its processing has a cost which must be balanced by the worth of the evidence bearing on the success or failure of the design.

(12) *Bases for Decision.* A design project (or subproject) is terminated whenever confidence in its failure is sufficient to warrant its abandonment, or is continued when confidence in an available design solution is high enough to warrant the commitment of resources necessary for the next phase.

(13) *Minimum Commitment.* In the solution of a design problem at any stage of the process, commitments which will fix future design decisions must not be made beyond what is necessary to execute the immediate solution. This will allow the maximum freedom in finding solutions to subproblems at the lower levels of design.

(14) *Communication.* A design is a description of an object and a prescription for its production; therefore, it will have existence to the extent that it is expressed in the available modes of communication.

Among the foregoing, there are principles of two kinds. Some are propositions that have a factual content.[1] They are factual because we can compare them with physical reality; thus their truth can be tested empirically. They describe what we believe is a proper generalization of some relevant part of physical reality. In the list, they are distinguished by the verb *is*. In terms of grammar they are in the indicative mood.

The remaining propositions have an ethical content. They reflect what we believe is a proper generalization of the values and mores of our culture. They are characterized by the auxiliary verb *must*. They are in the imperative mood, like the Ten Commandments; and like the Ten Commandments they can only be tested in a pragmatic sense. If people generally like the results, then we assume that the corresponding principles fit the ethics of our society. We might individually disagree with the particular ethics; if so, we have the right to seek, or to offer, the leadership that could persuade people to change them.

Berkeley, Calif.

NOTES

* First published in *Introduction to Design,* Englewood Cliffs, Prentice-Hall, 1962, Chapter I (=pp. 1–6).

[1] The notions of factual and ethical contents of statements are set forth lucidly in Chapter III, 'Fact and Value in Decision Making', of Herbert A. Simon's *Administrative Behavior*, The Macmillan Company, New York, 1958.

R. J. MCCRORY

THE DESIGN METHOD – A SCIENTIFIC APPROACH TO VALID DESIGN*

I. DEFINITION OF TERMS

Design. The process of selective application from the total spectrum of science and technology to attain a system, device, or process which serves a valuable purpose.

Methodology. The framework within which a sequence of steps can be based and from which check points to evaluate progress can be established.

Design Method. The methodology of design.

Stages of the Design Method. The levels of attainment within the framework of the Design Method.

Steps of the Design Method. The techniques and procedures used to progress from one stage to the next in the Design Method.

State of the Art. The current total of technical and scientific knowledge as obtained both from the literature and from personal and corporate experience.

Need. The motivation for design activity representing the expectation of payoff.

Design Concept. A general image of the system, device, or process constituting a design approach which has sufficient potential of satisfying the defined need within the state of the art to justify proceeding to the next step in the Design Method, the evaluation of feasibility.

Synthesis. The intellectual act of trying out various combinations of technological capabilities to evolve a design concept.

Design Feasibility. The confirmation that, in detail form, the design concept is attractive in terms of satisfying the need and being technically attainable.

II. INTRODUCTION

In our preoccupation with the details of design technology, it is easy to lose sight of why we are working on particular design tasks and how these tasks fit into the total role of the designer. This paper is an attempt to interrelate the facets of the overall design process and to establish that there is a methodology to design which can be a useful reference in defining where we are, where we ought to be, and how we should proceed in executing a design program.

Design is regarded as the process of selectively applying the total spectrum of science and technology to the attainment of an end result which serves a valuable purpose. It is the segment of engineering which devises and develops new things, in contrast to other segments which emphasize the solving of problems or the generation of engineering information. The responsibility of the design engineer is to use the maximum powers of creativity, judgment, technical perception, economic awareness, and analytical logic to devise uniquely useful systems, devices, and processes. His function is usually not to originate the basic scientific building blocks, but rather to utilize them so that the result is a useful creation.

The designer could be considered somewhat similar to the artist. The artist does not create new colors and forms. He combines colors and forms into new creations; and the results are, at times, masterpieces.

But imposed upon technological design is a crucial requirement which has no counterpart in classic art. Design must adhere to a plan which has objectives involving cost, performance, effort for attainment, probability of success, and even aesthetics. The fact that design must traverse a closely evaluated path, starting from a well-considered if not urgent need statement to a functioning achievement, requires that it follow a methodology. By methodology is not meant the tricks of the trade such as drafting competence, or analytical ability, or for that matter a flair for brainstorming. Methodology in design is considered rather as being the framework for the design process within which a sequence of action steps can be based and from which check points to evaluate progress can be established.

The design process being guided by and evaluated in terms of a method is comparable to the scientific community's use of the 'scientific method'. Although many scientists discount the conscious use of the scientific method, reflection upon scientific progress shows that it is applied and that scientific 'laws' have been accepted or rejected very much in accordance with the criteria demanded by the scientific method.

III. REVIEW OF SCIENTIFIC METHOD

Figure 1 is a model of the 'scientific method' which might be used as a starting point for describing a 'design method'. Ideally, the sequence of the scientific method begins with the body of (1) *existing knowledge* as represented by accepted scientific laws. As the scientist creatively observes

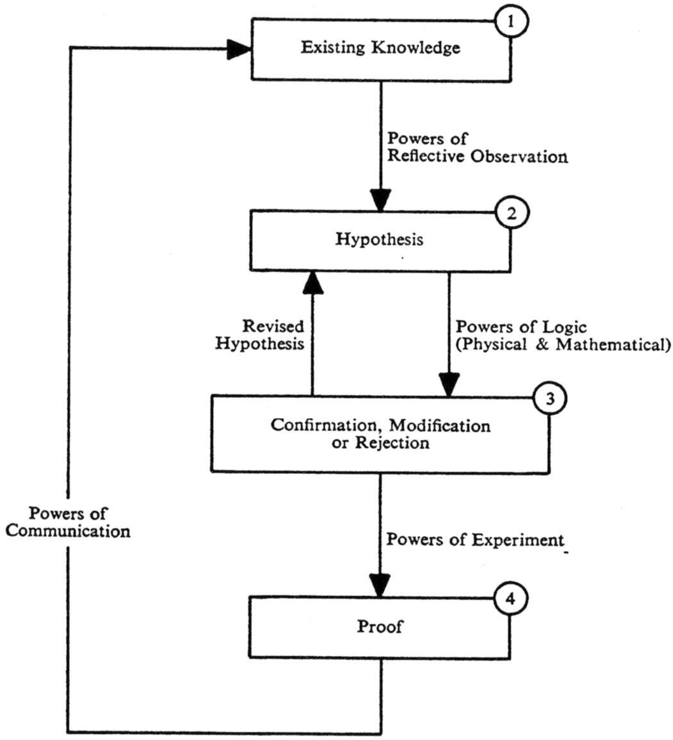

Fig. 1. Graphical representation of scientific method.

nature in the light of existing laws, he conceives of a (2) *hypothesis* which might explain some phenomenon. But this hypothesis is only the seed; it must be enlarged by logical analysis until it can be (3) *confirmed, modified, or rejected.* At any point, the scientific method may lead the scientist back to the hypothesis stage once again. But even if the hypothesis is confirmed in the eyes of the originator, he must have his confirmation accepted as (4) *proof* by his peers. This he does by having his contention tested under a variety of situations until it can be accepted as a truism within carefully defined boundary conditions. But even at this stage the full course of the scientific method has not been traversed. The loop must be closed with communication of the new scientific fact to the scientific community to enlarge the body of (1) *existing knowledge* from which other researchers may derive new hypotheses.

This idealized version of the scientific method is usually not categorically and sequentially used by scientists. There are many simultaneous activities going on which involve doubling back to check and revise and jumping ahead in anticipation of results in order to originate new hypotheses. Nevertheless, the scientific method is so inherent in guiding and evaluating scientific progress that it relentlessly imposes itself regardless of short-term abandonment.

IV. STRUCTURE OF THE DESIGN METHOD

The design process follows a methodology similar to that of the scientific method, although the design method has not been as carefully defined nor historically as well established. Nevertheless, the design method is as inherent to the design process as the scientific method is to scientific exploration. Designers will do well to recognize its structure so that the design method can be used consciously to clarify some of the costly 'mysteries' of design.

The design method, graphically described in Figure 2, parallels the scientific method. Each is a closed loop, with experiences gained during the execution and completion of the process providing the basis for subsequent applications of the method. Both the scientific method and the design method have stages, defined plateaus which can be used to identify when a recognizable degree of attainment has been reached. The comparable stages of the two methods are given in Table I.

To progress from one stage to the next, each method requires that certain intellectual powers be brought to bear. For the two methods, Table II gives the comparable steps that are involved.

The design method is compounded by multiple interconnecting steps and auxiliary stages. But these are ancillary to the basic methodology and can vary depending upon individual situations.

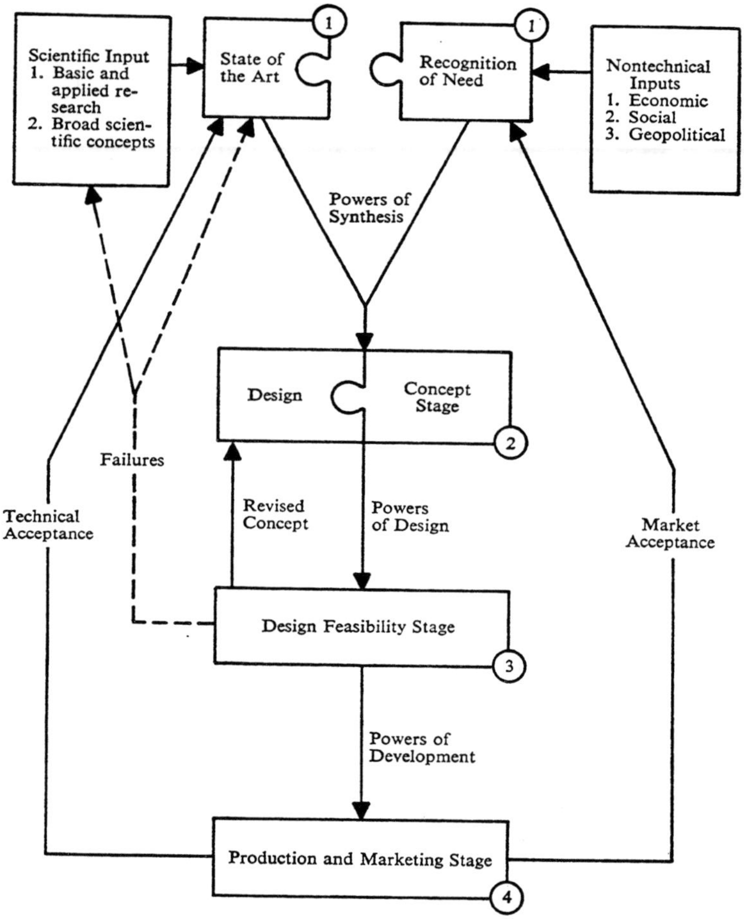

Fig. 2. Graphical representation of design method.

1. *State of the Art and Recognition of Need*

The starting point of the design method is more comprehensive than that of the scientific method. Unlike fundamental scientific research, design is motivated by need rather than by curiosity. Therefore, in addition to requiring knowledge of the state of the technical art, the design method requires recognition of a need which warrants an investment of effort and funds.

Recognition of need can be considered the marketing input to the design method. In the case of industry, this input involves the analysis of market needs in the light of corporate objectives. Military and space programs are planned to meet needs identified through broad studies of our overall security posture and anticipation of future move-countermove situations.

But whether government or industrial needs are being considered, the designer must realize that much of the input required to define the need is not technical, but rather socio-economic-geopolitical. Therefore, the designer must appreciate those key nontechnical factors which are significant in defining whether the results of his design work will fulfil a basic social, economic, or security need. With this appreciation, the designer is better qualified to extrapolate current requirements and creatively anticipate tomorrow's needs.

TABLE I

Stage	Scientific method	Design method
1	Existing knowledge	State of the art/recognition of need
2	Hypothesis	Design concept
3	Confirmation	Design feasibility
4	Proof	Production and marketing
1	Existing knowledge	State of the art/recognition of need

TABLE II

Steps between stages	Scientific method	Design method
1–2	Powers of reflective observation	Powers of synthesis
2–3	Powers of logic	Powers of design
3–4	Powers of experiment	Powers of development
4–1	Powers of communication	Powers of acceptance

Given a need-oriented assignment, the designer can encounter a series point of vulnerability; he must realize that the purpose of the design effort is to produce something which is truly useful. The purpose is not, as he might prefer, to provide a result which is technically self-gratifying and elegant, but from which the payoff can only be a technical paper to his peers. Because designers are required to satisfy profit or security motives, the definition of need is critical to the design method and each succeeding stage must be planned and judged on the basis of the need.

Along with the recognition and definition of the need, the design method requires an appraisal of the pertinent state of the art. State of the art includes materials capabilities, phenomena understanding, and previous design experience. As important as previous experience is, if designers are unduly dependent upon experience, design progress can be reduced to a series of small improvements. The design method requires that the designer tap into the total spectrum of technology with the objective of obtaining the greatest design advance consistent with the state of the art, wherever the art may exist. Experience which exists in technical or product fields foreign to that of the designer can often suggest the most advantageous design approaches.

But perhaps even more significant to design advances is the input which can be obtained from basic and applied research. As materials are devised and phenomena quantified, new raw inputs to the design method are made available. New scientific concepts such as nuclear heat sources and semiconductors open up vast areas for design exploitation. The design method, therefore, necessitates keeping open a direct link between the resources of scientific research and the state of the art available to the designer. The degree to which the designer can intercept the latest scientific information can determine the extent to which he can make significant design advances.

2. *The Design Concept*

A design concept is created when, through the designer's powers of synthesis, a recognized need and technical capability as represented by the state of the art are matched. When the designer can arrange technical art into useful combinations which form a system satisfying a need, he has a design concept.

Matching can originate from either the need or the art. Given a defined need, the designer can search the art for the inputs which can be synthesized to satisfy the need. Conversely, there are many concepts which are originated largely on the basis of known art, and the concept stage is attained by finding a need which can be fulfilled. The latter approach to design conception is the principal justification for the massive engineering research being conducted in energy conversion, materials, and other generic fields of technology.

If an idea does not satisfy a need, a design concept as defined by the design method does not exist regardless of how clever or novel the idea might be. Nor does a concept exist when a design which would satisfy a need requires a capability beyond the state of the art. The principal advantage to be derived is a feedback to the research laboratories.

The designer fulfils his synthesis function by an orderly procedure.[1] He first analyzes the need in considerable depth, perhaps allowing the need analysis to suggest a design concept. He then spatially visualizes systems which are advantageous combinations. Or he might utilize a further technique which is not as broadly recognized. This is to explore analytically the area of design interest, manipulating generalized mathematical expressions with the hope that unique design approaches will be derived which would not be apparent from only spatial analysis.

Attainment of the design concept stage of the design method means that a design approach has been derived having the potential of satisfying the need as well as the potential of being attained without violating the state of the art. Many ideas may be rejected by the designer before he arrives at one which qualifies as a design concept. On the other hand, he may finally have available more than one design concept showing attractive potential. At the design concept stage, the concept need not be described completely. Rather, it may be expressed in terms of functional requirements or 'black boxes'. The key criterion is that, in the judgment of the designer, the concept has sufficient potential to justify further effort in designing the individual elements of the system.

3. *Design Feasibility*

The design-concept stage having been attained, the next step of the design method is to establish design feasibility. Feasibility is established by determining whether all of the necessary functions of the system can be worked out and whether, when the design is in detail form, it still is attractive in terms of the need and the probability of successful attainment. To go from the design-concept stage to the feasibility stage of the design method means to convert the design as described in its functional form to specific elements. The steps[2] which may be used in this conversion process include:

(1) Definition of the concept in terms of its optimum combination of functions.

(2) Expression of detail design requirements in terms of functional and/or performance specifications.

(3) Design of specific elements to meet specifications (to be done in accordance with design method using specifications as need statement).

(4) Tradeoff analysis comparing design alternatives and, if required, revision of specifications.

(5) Critical experimentation to test specific questionable aspects of the design concept.

(6) Operation of experimental prototype to confirm adequate functioning of total system or subsystems.

Frequently the design concept fails to reach the stage of feasibility because the technical problems cannot be solved successfully or because the concept does not fulfil its apparent potential of being attractive in terms of the need. The probability of this happening should be reduced if designers knowledgeably go through the prior steps of the design method. When failure does occur, it is necessary to return to the concept stage and revise the concept in the light of the intervening experience. However, if failure is complete, the most that can be salvaged is the failure experience which can be interjected into the state of the art as a guide to subsequent design programs.

4. *Production and Marketing*

When the designer is convinced that a feasible design is in hand, he is ready to move on to the next stage in the design method. This is the attainment of a design which can be produced successfully and marketed. The process by which a feasible design is converted into a production design is broadly termed development. In actual practice, development tasks are extremely demanding in terms of engineering skills and usually involve the major expenditure of funds and time in the entire design process.

Within the framework of the design method, the development step is still very much a design function.[3] The designer remains responsible for perfecting the design in terms of performance, reliability, and cost. Although specialists in value analysis, tooling, field testing, and marketing may more prominently enter the picture, they cannot recover success if the requirements of the earlier stages of the design method were not validly satisfied. Presuming skillful engineering and marketing, the development steps, although costly, are not highly risky. The mistakes which lead to disastrous failures are more likely to occur at the stages when decisions regarding need, concept selection, and feasibility are made.

The loop of the design method closes when the design is judged a technical and marketing success. This experience in market acceptance extends the understanding of need and invariably leads to the identification of new areas of need. The technical successes and failures expand and temper the state of technical art and are inserted into other design programs.

V. VARIANCES TO THE IDEALIZED DESIGN METHOD

The design method is presented in the foregoing as an idealized representation of the design process. If individual design programs were discrete entities, the design method as described here in skeletal form would be accurate. In actual fact, design programs seem to blend into one another to the extent that a complete progression of the design method is difficult to recognize. As actually practiced, design progress involves a pattern of superimposed programs, each subject to the requirements of the design method. This pattern may include:

(1) Design programs in which the outcome is the selection of a means of approaching a broad national or industry need. This type of program is being applied to the transportation problems of the East Coast Megapolis. The application of the design method here will involve overall systems concepts whose feasibility will be established by computer analysis rather than by operation of prototype hardware. The outcome will be the need statements for other design programs.

(2) Numerous design programs which originate from the same need recognition. Some will be parallel programs conducted by different design groups searching for the same end result. Theoretically, application of the design method should result in the same outcome for each program. But, the personalities and backgrounds of the individual design groups are so important to the functioning of the design method that results which are different in both approach and quality are inevitable.

(3) New programs which are initiated before an original design pro-

gram is completed which have as their objectives the improvement upon the results of the first program. The demands of progress often cannot wait for the sequential completions of related design programs.

(4) Auxiliary programs on subsystems and components which are part of an original overall program. The concept having most potential may require an element which is identified as beyond the state of the design art. But the attractiveness of the concept warrants the risk of initiating a separate design program to expand the state of the art. While establishing feasibility, the designing of individual functional members becomes a number of subprograms calling for use of the Design Method.

VI. AIRCRAFT-GAS-TURBINE DESIGN HISTORY AS AN EXAMPLE OF THE DESIGN METHOD

A general review of aircraft-gas-turbine history can help demonstrate the applicability of the design method.[4] Although the modern development of the gas turbine was accomplished without benefit of a unified and acknowledged design methodology, in retrospect it must be recognized that the present state of the art was reached by a route strikingly similar to the design method described in the foregoing.

Figure 3 is greatly simplified graphical representation of the aircraft-gas-turbine development as described in terms of the design method. Although there was considerable gas-turbine history before the 1930's, the 'breakthrough' concepts came about when state of the art and need matched. The state of the art consisted of experience from earlier research by Stodola and others, knowledge of the properties of improved materials developed for turbines and internal-combustion engines, and know-how relative to centrifugal compressors derived from work on aircraft turbochargers.

Both the British and the Germans had a prime need motivation in their search for aircraft propulsion systems having markedly higher potential for performance than those of their enemies. The need and the art were matched when it was recognized that the synthesis of the centrifugal compressor and materials permitting higher turbine-inlet temperatures made possible a concept which might fill the urgent need. Although German efforts soon switched to work with axial compressors, the British investigation of the new concept brought to birth the modern aircraft gas-turbine.

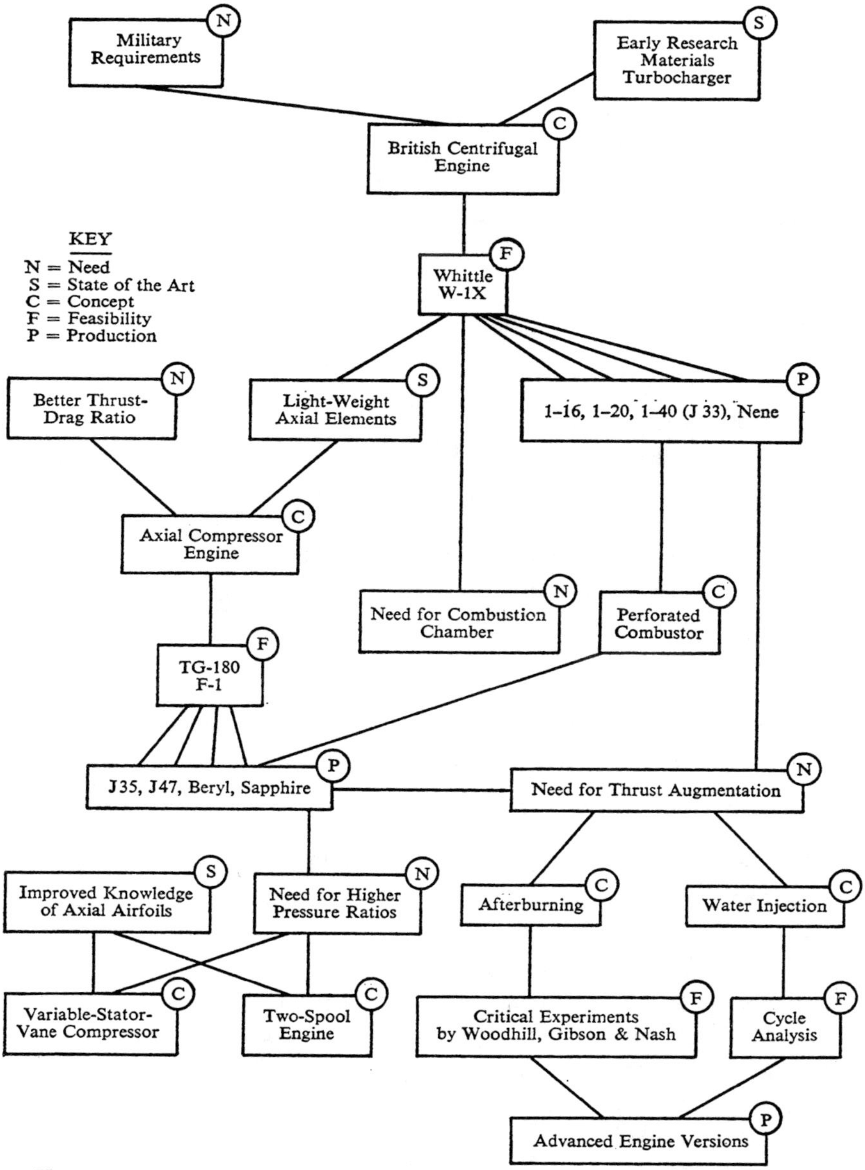

Fig. 3. Graphical representation of design method as superimposed upon aircraft gas-turbine history.

Feasibility of the new concept was confirmed by Whittle's W-IX engine which was developed into a family of operational engines culminating in the J33 in the United States and the Nene in England. When centri-fugal-compressor engines became operational and experience gained became part of the state of the art, the design method circuit was completed.

However, long before that cycle could be completed, other cycles of the design-method application were set in action. They were based upon (1) new engine concepts suggested by improved recognition of need and by new art being continually developed, and (2) new subsystem concepts required to provide critical inputs to the success of engine concepts.

As an example of the necessity of devising subsystem concepts, experience with the earliest engines demonstrated the critical need for a better combustion chamber. This urgent need led to the concept of a perforated combustor which introduced a protective air shield and increased combustion turbulence. Feasibility was established by critical experiments on early engines, and the nearly universal use of perforated combustors on all gas turbines provided proof in use. Therefore, the design-method cycle was again completed. A similar design cycle involving many subsystems and components is continuing. For example, at the present time the design method is being vigorously applied to new concepts for aerospace-engine fluid fittings.

Even as the work on the basic Whittle gas-turbine concept proceeded through the steps of the design method, new or modified gas-turbine-concepts were being pursued and succeeding or failing on the basis of the criteria of the design method. Two that have not succeeded because of the uncompromising standards of need or available technology have been the pulse jet and the piston engine acting as a gas generator for an aircraft turbine.

Other concepts have successfully followed and have superimposed themselves upon the development of the basic gas-turbine concept. When the need for a more slender frontal profile was recognized and when the experience with lightweight axial turbines in the early engines suggested lightweight designs for axial-flow compressors, the axial-compressor-engine concept gained favor. Its feasibility was established with the prototype TG-180 and F-1 engines in the United States and Britain. These in turn led to the operational family of axial-flow engines represented by the J35 and J47 and the Sapphire.

Concurrently, the need for thrust augmentation led to concepts of afterburning and water injection. Feasibility of afterburning was established by critical experiments showing that additional fuel could be burned in the exhaust from gas turbines. A cycle analysis was instrumental in showing feasibility for water injection. The development of advanced engine versions utilizing both injection and afterburning completed these cycles of the design method. These concepts were soon followed by two-spool machines and many other concepts which continue to be evolved as need and state of the art are matched.

VII. APPLICABILITY OF DESIGN METHOD

The design method as outlined is not to be considered as restricted to large programs having comprehensive objectives. It has been and is being used on programs having objectives as varied as hair dryers, hydraulic systems, large radio antennas, and advanced nuclear-reactor systems.

Varying degrees of management may be required depending upon the nature of the individual programs. For vast systems programs, the design method can be administered through formalized procedures. For smaller programs, the designer himself may monitor the execution of the program to conform to the requirements of the design method. Here the attitude and comprehension of the individual designer is critical to the adoption of the design method.

Whether the design method is applied to a design program which is large or small, its basic purpose is the same. The design method is to serve as a disciplining influence in providing a framework upon which design progress can be planned and design pertinence evaluated. In order to proceed in accordance with the design method, the following conditions must be satisfied sequentially:

(1) There must be a well-understood starting point for the design in terms of:

(a) A thorough awareness of the supporting technology.

(b) A clear definition of the objective. These considerations are so important that considerable technical and marketing research can be justified to attain the required levels of understanding.

(2) The concept must be the result of a searching synthesis based upon (1). Many potential concepts should be obtained and scanned for pre-

liminary determination of general feasibility and attractiveness. The design method requires searching for the best obtainable concept rather than accepting the first evolved, and it should be a restraint against premature selection.

(3) Feasibility must be established in terms of whether:

(a) The concept retains its pertinence to the need.

(b) All facets of the concept have been considered and are acceptable.

(c) The remaining steps are realistically gaged.

The feasibility stage must be approached coldly to avoid over-optimism in presuming feasibility and to ensure realistic estimating of the effort required to reach the next stage of production and marketing success.

In addition to being a controlling influence upon the design process, the design method also serves as a communications medium. Management often does not understand the steps through which a design program must pass between its inception and completion. Consequently, miscalculations can result which cause either the continued support of unpromising programs or the premature cancellation of some which are promising. A typical example of misunderstanding is the false assumption that a prototype which only shows feasibility is a production model. The design method can provide a 'universal language' understandable to both the designer and his management. Identification of program status as being at or between stages of the design method can provide the communications link to permit management to conduct surveillance and planning realistically with fewer problems of semantics.

VIII. CONCLUSION

The purpose of this discussion is to identify and describe a methodology which should be inherent in all design programs but which is seldom recognized and consciously utilized. The failures resulting from attempts to avoid or shortcut the stages of the design process can be reduced if the controls and guidlines of the design method are knowingly used. In practice, the design process as exemplified by the design method is often badly distorted, especially in the critical early stages. Need is seldom adequately defined and the socio-economic factors are often miscalculated. The route by which the designer may tap the state of the technical art is long and hazardous, and the eagerness to accept and invest heavily in initial ideas sometimes causes designers to overlook the importance of the concept synthesis steps. In many programs the feasibility stage is by-passed.

The idealized Design Method presented here should be adapted to the peculiar requirements of each design task. Then it should be used as the disciplining influence and communicating tool which is needed to make design a more profitable and satisfying endeavor.

Columbus, Ohio

NOTES

* First published as *ASME* (American Society of Mechanical Engineers) *Paper*, No. *63-MD-4* (1963).

[1] R. J. McCrory, W. H. Wilkinson, and D. W. Frink, 'Synthesis of Concepts in the Design Method', paper presented at ASME Design Engineering Conference, New York, N.Y., May, 1963.

[2] H. A. Cress, and E. S. Cheaney, 'Determining Design Feasibility', paper presented at ASME Design Engineering Conference, New York, N.Y., May, 1963.

[3] R. J. McCrory, 'The Science of Design', paper presented at Second Conference on Engineering Design Education, Los Angeles, Calif., September, 1962.

[4] I. H. Driggs, and O. E. Lancaster, 'Gas-Turbine Development – Aviation', ASME Paper No. 52–F-7, September, 1952.

A. D. HALL

THREE-DIMENSIONAL MORPHOLOGY OF SYSTEMS ENGINEERING*

I. INTRODUCTION

Morphology refers to the study of structure and form. Morphological analysis, a term coined by Zwicky [1], means to decompose a general problem or system into its basic variables, each variable becoming a dimension on a morphological box. When the values that each variable can assume are found, a set consisting of one value of each variable defines a solution to the problem or a species of the general system. This valuable approach is essentially a search technique for piling up alternatives in a design problem [2]. In this paper the technique will be used to present a new and simple model of the field of systems engineering that may be useful in surprising ways.

II. MORPHOLOGICAL ANALYSIS

Investigation of systems engineering reveals (at least) three fundamental dimensions which are as follows.

(1) The first is a time dimension which is segmented by major decision milestones. The intervals between these milestones can be called phases, and they define a coarse structure depicting a sequence of activities in the life of a project from inception to retirement.

(2) The second dimension models a problem solving procedure, the steps of which may be performed in any order, but each of which must be performed no matter what the problem. These steps may be repeated in successive phases. The flow of logic, not time, is the essential feature of this dimension, and this logic comprises the fine structure of systems engineering.

(3) The third dimension refers to the body of facts, models, and procedures which define a discipline, profession, or technology. A possible measure for this dimension is the degree of formal or mathematical structure. The intervals along this scale in decreasing order of formal structure

might be: engineering, medicine, architecture, business, management, law, the social 'sciences', and the arts.

Combining the first two dimensions produces a model of the methodology of systems engineering which at once organizes and defines the field independent of any profession. This does, of course, imply that systems engineering is not a profession, since it does not contain facts, models, technology, etc., that are unique to it. To the extent that systems science is succeeding in abstracting models, concepts, facts, etc. that apply to several fields, it is fair to characterize systems engineering as an emerging profession. Figure 1 depicts this model, called the activity matrix because each element in the matrix is defined by a unique activity at the intersection of a phase and a step of that phase. The model encompasses a vast panorama including design, which is centered about phases 2 and 3.

Most of the two-dimensional structure has been discussed previously [3]. Therefore, these broad phases and steps will be defined only briefly. By program planning is meant conscious activity in which an organization strives to discover the kinds of activities and projects it wants to pursue into more detailed levels of planning. In the language of finance, it is portfolio design.

Project planning is distinguished from program planning by interest focused on just one project of the overall program. The terminal milestone of this phase occurs when a decision is reached to develop the best of the alternative systems disclosed during the planning or to dispose of the project in some definite way.

This or any phase can be defined in terms of the steps which comprise it. Thus problem definition, activity a_{21}, includes a study of the needs and environment, and collection and analysis of data to be used in formulating the problem. Value system design, activity a_{22}, uses these data in stating the objectives to be met and prescribing a (generally multidimensional) decision criterion against which all alternatives will be measured. System synthesis refers to all means of compiling a set of contending alternatives, whose consequences are systematically deduced during the systems analysis step. These consequences are evaluated and combined according to the rules prescribed by the value system in the decision making step, which selects the best alternative. Rational choice among alternative systems requires that each system be proportionated to meet, as best it can, the objectives comprising the value system; this is the role of optimization.

Steps of the Fine Structure (Logic →) / Phases of the Coarse Structure (Time ↓)		1	2	3	4	5	6	7
		Problem Definition	Value System Design (develop objectives & criterion)	Systems Synthesis (collect & invent alternatives)	Systems Analysis (deduce consequences of alternatives)	Optimization of each Alternative (iteration of steps 1–4 plus modeling)	Decision Making (application of value system)	Planning for Action (to implement next phase)
1	Program Planning	a_{11}	a_{12}				a_{16}	a_{17}
2	Project Planning (and preliminary design)	a_{21}						
3	System Development (implement project plan)							a_{37}
4	Production (or construction)				a_{44}			
5	Distribution (and phase in)							
6	Operation (or consumption)	a_{61}						
7	Retirement (and phase out)	a_{71}	a_{72}				a_{76}	a_{77}

Fig. 1. Morphology of systems engineering and its activity matrix.

In the sense that this step entails iteration of the first four steps, it should not be singled out as a separate function. However, optimization often carries out this iteration by using a model for selected aspects of the system with the express purpose of optimally proportioning the selected aspects. It is this modeling activity which justifies singling it out as a separate step. The final step is planning for action, which includes communicating results, scheduling effort, allocating resources, determining how performance is to be measured against the plan, and designing a feedback system for controlling the ensuing action. Were we not modeling a multiphasic system, we would include implementation, i.e., starting and controlling action, as a final step. However, in this model, implementation refers to the next phase.

Thus system development means to implement the plan. It entails another cycle of steps, dealing mostly with components rather than overall alternatives. The phase ends by preparing detailed specifications, drawings, and bills of materials for the manufacturer or construction organization.

Production, in the case of a manufactured product, or construction, when the system must be produced in place, refers to all those activities needed to give physical embodiment to the wanted system. For a new building, the general contractor executes the architect's plan, using the detailed plans and specifications provided by him and his consultants. For a new product, the manufacturing engineers determine the sequence, material flow, and the floor layout required, design the tooling and test jigs, and establish quality control.

Next follows distribution and phase-in of the product to ultimate consumers. This may involve all kinds of distribution facilities, sales organizations, applications, and sales engineering. The product may have a very long life, like a power dam, or it may be consumed, like a new item of packaged food.

The operations phase overlaps the distribution and retirement phases a little or a lot, depending upon the number of systems involved and the periods used for phase-in, operation, and phase-out. In any case, operation is the reason for all forms of systems engineering. Many problems arise during this phase that are not of a design nature, such as those relating to optimum utilization, which are solved by a recycling of the seven major steps of the fine structure.

Finally, the system may be retired, or more generally, phased out over a period of time while some new system takes its place. Just as for all of the other phases, a whole row of steps applies.

Consider now that a matrix of 49 activities is formed by the coarse and fine structure dimensions.

The activities of each morphological square are unique, yet there are helpful similarities and relationships. For example, the objectives selected for a particular value system design may differ according to which phase we are in; a type of objective appropriate in program planning more than likely would be inappropriate, even irrelevant, for the retirement phase. Yet in-depth knowledge of how to design and apply value systems is useful in all phases and can lead to wisdom in tailoring a value system to a phase.

Modeling the fine structure as a linear dimension on a box over-emphasizes the temporal features of the fine structure and obscures certain essential features of the systems engineering process. Both the iterative and converging features of the process are emphasized by using a simple natural system as a model (thus using the essential cybernetic viewpoint [4]): a seashell viewed as a cornucopia with reverse flow. See Figure 2.

The analogy of the cornucopia with the systems engineering process is a felicitous one. The cornucopia, emblematic of abundance, was the horn of the Greek nymph Almathea which was endowed with the virtue of becoming filled with whatever its possessor wished. Here, the wishes are those of a society directing and cooperating with its systems engineers, who focus energy, information, and materials upon a successively smaller and smaller set of problems and decisions until finally a single wanted system emerges and is fit into an ecological niche. The spiraling structure converging to a point depicts exactly what happens in iterating the fine structure cycle through successive phases.

All dimensions of the spiral horn are adaptable to fit the task (as required of any good cornucopia). If we stretch the radial and transverse axes to separate the segments (taking care not to go beyond the elastic limit) and look into the larger end, we may see the hyperfine structure of this remarkable instrument. What we see is a series connection of one-way elements, providing the structural basis for forward movement, connected with two-way (feedback and feedforward) paths permitting any step to follow any other step in whatever sequence.

Seemingly complex beyond belief or comprehension, because we know how intractable even a single-loop feed-back system can be if it is of high

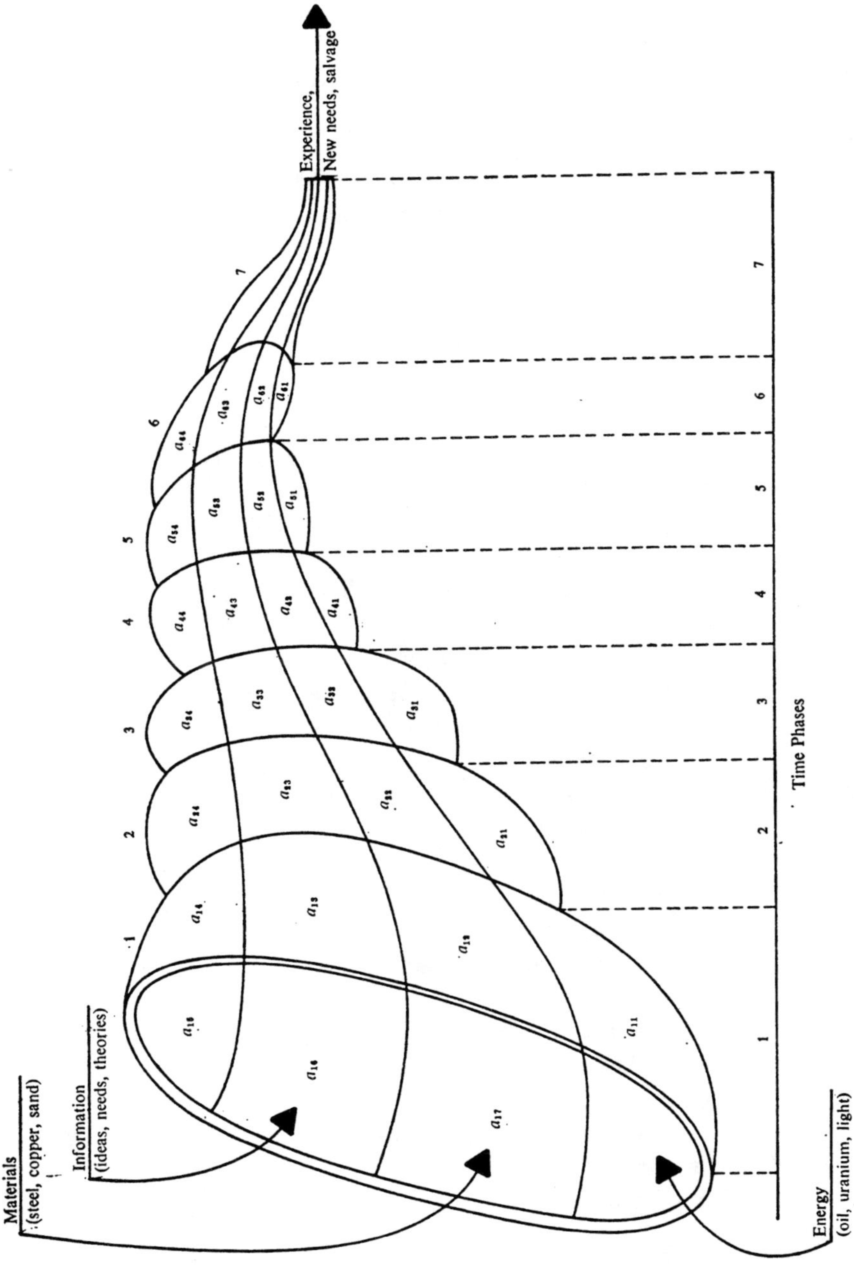

Fig. 2. Cornucopia model.

order and nonlinear as those are in this model, certain regularities do appear. At least, certain feedback paths appear 'stronger', 'more essential', or perhaps only more frequently used than others. The loop consisting of analysis, synthesis, and comparison with objectives in each phase has been singled out by some methodologists. Special emphasis has been given to this loop here also, and the loop which models value system design has been added and connected with the synthesis-analysis loop [3, Figure 4.3]. How these two loops behave like an adaptive feedback system has been explained. Although progress has been made, Figure 3 shows how far we still have to go to achieve complete understanding.

It must be perfectly clear that the two-dimensional morphology presented is quite different when applied to a problem in electrical communications than it is when applied to a problem in medicine or bridge construction. It follows that even if a man were perfectly versed in the tools, models, and attitudes appropriate for all of the activities on the two-dimensional morphology matrix, this would not be sufficient to produce anything really comprehensive and useful in the real world where specific knowledge and technology must be applied. The requirement for subject matter knowledge in practice should be self-evident, but, unfortunately, it is not always so. As evidence, we see consultants in operations research and industrial engineering who tend to claim that since they have grasped a universal methodology, they can produce applications in any field. Also, certain aerospace companies, claiming the systems approach, have tackled problems in transportation, education, and pollution with results that may not be what one would expect after contemplating cornucopias. Even many universities, uncritically being swept along by the tide, have established graduate curricula in systems engineering that cover a good part of the methodology, but leave the student unequipped to design anything. A few are extending the same sort of programs downward into undergraduate curricula.

Thus are we led to the third dimension, referring to subject matter fields representing what today are called professions, disciplines, or technologies. Figure 4 defines a more well-rounded systems engineering. Within it, we may speak of more usefully defined activities such as decision making in the development phase of law (a_{365}) or the operational phase of medicine (a_{612} through a_{672}), which includes the work of the general practitioner of medicine.

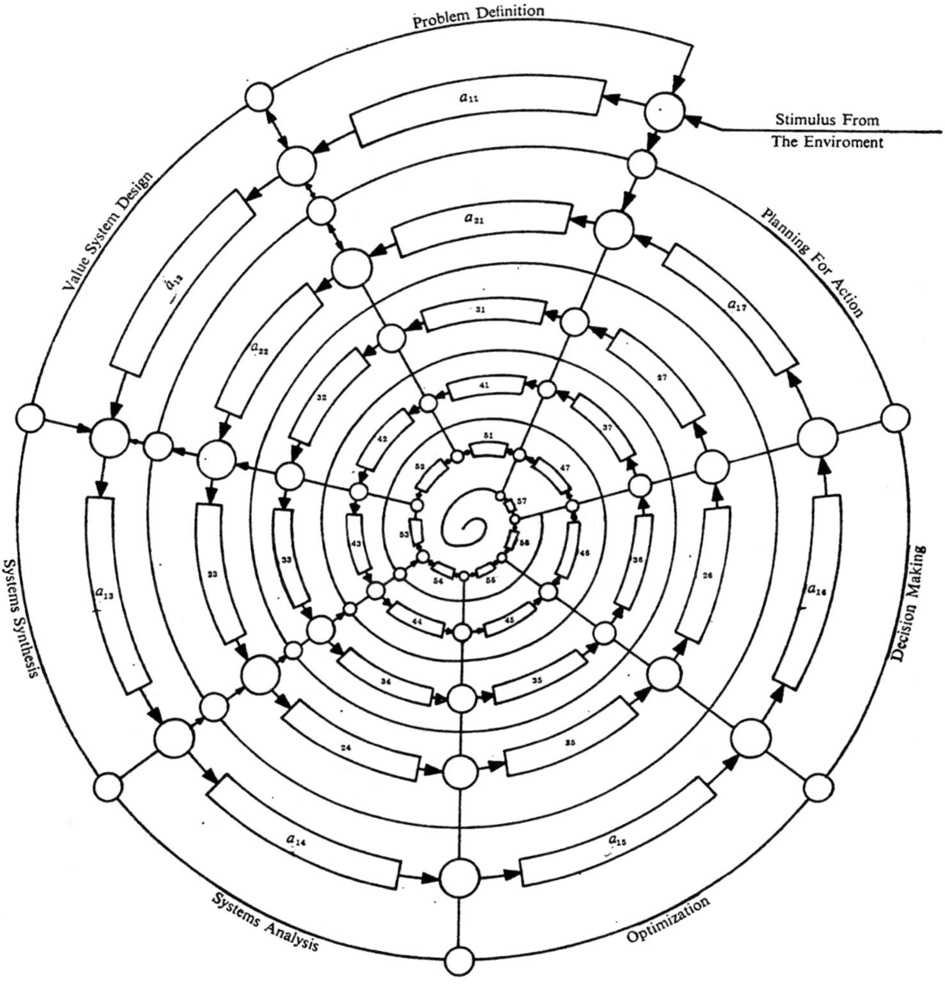

Fig. 3. Hyperfine structure.

III. APPLICATIONS

This morphological box appears to have many uses of which only three will be mentioned briefly.

A. *Taxonomic Uses*

As illustrated previously, it is possible to define well-known fields by reference to a set of compartments in the box. Attempts to do this lead to some problems caused by misnamed fields. As an example, consider the field known as 'systems analysis', which was started at the RAND Corporation and continued by a rather broad group of mathematicians and economists who, among many other things, performed military 'cost-effectiveness' studies. The box defines systems analysis as a deductive step only, occurring in all phases of all disciplines; this captures only part of systems analysis and credits it with working in fields where it is not found in fact. A better match occurs if we classify it as a part of the field of operations research.

Operations research started through interest in the operations phase, but it greatly expanded by using the steps in the fine structure to solve problems in other phases as well. It has spread over most of the 'zero plane' of the box and has found applications in common activities in other 'layers' of the box, notably, business, management, and medicine. However, in the sense that it does not use 'research methodology', meaning the scientific method as used in pure science, it too is misnamed. Its methodology seeks a normative or prescriptive body of knowledge, unlike pure science which seeks 'to know', and does not seek 'what ought to be'. The same applies to the 'science' of management science. Ansoff and Brandenberg [5] think that the right solution to this little taxonomic problem is to call it 'management engineering'. By the same arguments, operations research should become 'operations engineering', and 'systems science' should become 'systems engineering'. Thus a broad viewpoint contributed by the morphological box can help to improve tangled terminology.

B. *Aids to Discovering, or Seeing More Clearly, Unique Activities*

One of the distinct merits of morphological analysis is that it helps to find more solutions than could be found merely by listing them. In this case, the box identifies $7 \times 7 \times 8 = 392$ activities, and this number can readily be increased by anyone adding professions and subdividing phases and steps, according to his own viewpoint. The importance of this is that the delineation of many unique activities by the combination of variables invites a large number of new questions. (Any good research raises more questions than it answers.) One class of such questions is about the existence of a given combination. Many inventions have taken place by elaborating a combination which at first seemed foolish or farfetched. This class of questions will be exemplified.

Is each layer of the box really complete all over the plane? In other words, does each profession really have a methodology as well developed as 'conventional' systems engineering from which the box was derived? It is, of course, not too difficult to imagine problems in law, the arts, etc. and 'force fit' the methodology of the first layer to them, but this proves nothing, except perhaps our ingenuity.

There is very little evidence to suggest that a 'medical systems methodology' or a 'legal systems methodology' has been worked out. Nadler [6] reports interviews with an engineer, a lawyer, a commercial artist, a physician, and an architect about the procedure each followed in a recent specific design project. These were the results:

> The research group concluded that there were strong similarities in steps used by those interviewed. Although there was not enough evidence to support a conclusion that the design approaches of various professions could be translated into one design model, such an assumption seemed justified as a working hypothesis.

Even if the hypothesis were validated (the author's own experiences both in engineering and nonengineering areas lend support), it is a far cry from here to well-reasoned and formulated methodologies which have consensus within the various professions.

Yet what a boon it would be if a common methodology were developed for the professions! Agreement merely upon common names for phases, steps, and structural elements alone would assist greatly in the transfer of knowledge among the professions. The potent idea of the 'portable concept' given by Linvill [7] would begin to realize its potential. The 'two-culture problem' discussed so much by Snow [9] is most acute between pairs of professions farthest apart on the 'professions' scale of Figure 4, precisely because of the most diverse methodologies and subject matters. Yet the author's experience is that people 'farthest away' from engineering,

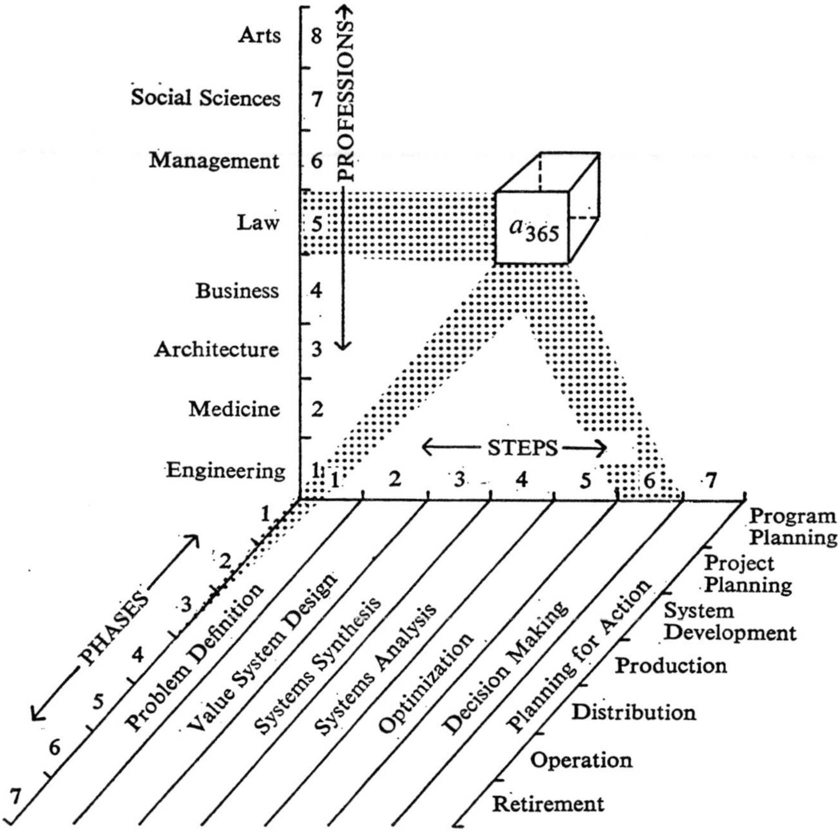

Fig. 4. Morphological box for systems engineering.

such as in the 'social sciences', have the greatest expectations that somehow engineering can give a great infusion of models, methodology, and techniques.

None of this should be interpreted to mean that engineering methodology is in such fine shape as to be a fit exporter. While being able to pinpoint its weaknesses, the existence of a tenable and useful methodology for engineering (including at least, systems engineering, management engineering, and operations engineering) seems beyond question, despite the more fainthearted proposition by Zwicky in [8] that one might now be attempted. There are, of course, plenty of scientists and a few engineers who gainsay all of this and who either hold to a purely heuristic method or fear that the evolution of a better methodology will lead to blind rule following. Both views are silly, and it would be illuminating to prod such people into print.

C. *Aid in Curriculum Design*

A final use for morphological analysis is in the design of academic curricula. In many new graduate school curricula in systems engineering, it is good to see emphasis on probability, statistics, mathematical optimization, statistical decision theory, etc. These tools are all useful and none should be banished. It is important, however, that these subjects be seen largely as the tools for the systems analysis step, which is only one of seven to be played together as an organ. The morphology suggests that there is no logical priority among the steps of a given phase; none of them is more important than the others. It follows that a unified curriculum requires some balance of emphasis upon each step in at least one phase of at least one profession. Schools which stress design courses achieve a better balance across the steps, but they give little balance across the phases since design centers about phases two and three.

The morphological box, of course, suggests tremendous challenges to curriculum designers to develop systems science and systems engineering which will permit one to operate effectively in several professions at once. Meeting these challenges calls for further weakening of the ossified partitions that separate the professions in most universities. False value systems, which tend to prevent research and teaching in fields that may not now be mathematizable, must also go. (Whoever heard of a course on problem definition?) In business, the rigid barriers between sales and marketing, research and development, manufacturing, finance, etc., are just as confining as those in the universities. And the unnatural wall between the mandarin society of the university and the pragmatic society of business serves neither in the evolution of systems curricula.

IV. CONCLUSION

We have to be modest even with perfect knowledge of a two-dimensional coarse and fine structure. But it is essential to see it whole to acquire a deep grasp of the process and facility with each of the activities of the process because it does, in fact, portray the essence of the systems approach. And this wholeness does aid in acquiring and using knowledge of disciplines comprising the third dimension.

Philadelphia, Pa.

NOTES

* First published in *IEEE Transactions of Science and Cybernetics* **SSC-5** (1969), 156–160.

BIBLIOGRAPHY

[1] F. Zwicky, 'Morphology of Propulsive Power', Society for Morphological Research, Pasadena, Calif., Monograph 1, 1962.
[2] S. A. Gregory (ed.), *The Design Method*, Plenum Press, New York, 1966.
[3] A. D. Hall, III, *A Methodology for Systems Engineering*, Van Nostrand, Princeton, N.J., 1962.
[4] W. R. Ashby, 'The Cybernetic Viewpoint', *IEEE Trans. Systems Science and Cybernetics* **SSC-2** (August 1966), 7–8.
[5] H. I. Ansoff and R. C. Brandenberg, 'A Program of Research in Business Planning', *Management Science* **13**, No. 6, 1967.
[6] G. Nadler, 'An Investigation of Design Methodology', *Management Science* **13**, No. 10, 1967.
[7] W. K. Linvill, 'Models and Model Construction', *IRE Trans. Education* **E-5**, (June 1962), 64–67.
[8] F. Zwicky and H. G. Wilson (eds.), *New Methods of Thought and Procedure*, Springer-Verlag, New York, 1967.
[9] C. P. Snow, *Two Cultures and a Second Look*, Cambridge University Press, London, 1964.

THE ROLE OF EXPERIMENTS IN APPLIED SCIENCE*

SIR – In a review of my book *Studies in Elastic Structures*, which appeared in your issue of August 14th, the reviewer questioned the role of experiments in applied science. He expressed doubt as to the propriety of raising such a general question in a review, but argued that the matter was fundamental to the philosophy of engineering science and called for discussion.

I should normally hesitate to enter into correspondence on a review of my own book, but because I agree so strongly with your reviewer as to the importance of this issue, I would like to develop my views somewhat in the hope of provoking the discussion he desires.

I have in an earlier book, *The Experimental Study of Structures*, suggested that experiments fall broadly into five groups, viz.:

(a) Exploratory experiments made before mathematical analysis.

(b) Confirmatory experiments made after analysis.

(c) Experiments made in conjunction with analysis to provide essential data or to obtain empirical formulas.

(d) Analytical experiments to replace computation in specific problems.

(e) Experiments of an *ad hoc* nature.

The last of these can be excluded from the present discussion since they would be classified by your reviewer as 'tests' and all engineers will agree as to their necessity.

In group (a) I include such experiments as those on the arch made of steel voussoirs, which are described in the book under review. These led to an understanding of the mechanics of voussoir arches and indicated a basis upon which such structures could be analysed. Later, experiments were made on larger arches with voussoirs made of two different concretes and jointed in two very different types of mortar. These, falling into group (b), were designed to show to what extent the methods based on the behaviour of the idealised structure could be applied to arches built more in accordance with actual practice. They provided information which could not, I believe, have been obtained by any other method and were thereby justified.

In the early years of engineering science it was common for large numbers of tests to be made on particular structural units and the results used to deduce empirical formulas for design purposes; the result of this method may be seen, for example, in the numerous formulas propounded for determining the strength of struts. It is not very satisfactory, and although it may prove convenient on occasion to summarise experience in this way, a far better approach is to study a problem as completely as possible by mathematical analysis and to fill the almost inevitable gaps by means of data obtained experimentally in a systematic manner. The numerous strut formulas just mentioned have, for example, been replaced in this country by the Perry-Robertson formula obtained exactly in this way. The original analytical treatment due to Professor John Perry contained a constant, depending upon the cross section of the strut and its equivalent initial curvature, the evaluation of which could only be made experimentally. As the result of an analysis by Professor Andrew Robertson of large numbers of tests from many sources, values were assigned to the constant which transformed an academic statement of strut behaviour into a valuable design formula which soon became the basis of British practice. This type of experimentation falls into group (c).

Experiments in group (d) are becoming increasingly important. Many engineers, for example, now realise that elaborate computation in the analysis of certain types of stiff-jointed frameworks can be short-circuited by simple displacement measurements made on small celluloid models, and whether such experiments are used as a substitute for the initial design calculations or simply as a completely independent check upon them, they are of great value not only for the actual results obtained, but because they lead to clearer understanding of the behaviour of the structures which they represent.

The doubt, expressed by your reviewer, as to whether it is worth while to reproduce in experiments the simplifying assumptions of the theory seems to indicate a misunderstanding of my position. It is true that in a very few instances experiments are mentioned which do no more than what he suggests, i.e. support computation because the assumptions of the analysis are accurately reproduced. As examples of this may be quoted a simple experiment on the deflections of a tightly stretched wire and an experimental check on the influence line of a bow girder. These experiments, however, were very simply and quickly made with existing apparatus, and while I agree that to the mathematician they are entirely superfluous, I have to confess that as an engineer I still derive pleasure in seeing such easily contrived experiments produce calculated results. More than this, on those occasions when the results differ I am grateful for the indication that something may have gone wrong. Very few experiments of this kind are mentioned, however, and perhaps reference to more elaborate experimental researches will serve best to explain my attitude. In connection with the airship programme in the years following the 1914–1918 war, new methods of analysis for braced tubular pin-jointed frameworks were based on certain assumptions as to the stiffness of bulkheads. The Airship Stressing Panel of the Air Ministry suggested the advisability of verifying these and experiments were accordingly made on braced hexagon tubular frames. The first frame was made to correspond as nearly as possible with the assumptions of the theory and the close agreement between the results obtained by calculation and experiment gave confidence in the experimental technique employed. If the work had stopped at this point I should be compelled to agree with the views of your reviewer that it was hardly worth while, but in fact this was only a beginning.

The structure was successively modified in different ways to cover cases outside the theoretical treatment, e.g. by the omission of bracing in certain bulkheads, by variation in the stiffness of such bracing, by fixing a nose piece on the tube to modify the resistance of the end bulkhead against distortion out of its plane, and so on. It was also used to determine the applicability of Saint-Venant's principle to a braced tube – a matter not only of considerable importance for the airship stressing scheme, but also one of scientific interest. Finally, a completely new model was made to determine the effect of continuous longitudinal members.

The second series of experiments to which I would refer were those on wire wheels, which well illustrate group (b). An early analysis had shown the impracticability of obtaining an accurate numerical result for the stresses in a wire wheel with more than six-spokes owing to the labour involved in solving ill-conditioned equations, and a new approach was made based on the assumption that a finite number of spokes could be replaced by an idealised disc of peculiar properties. The treatment of this idealised structure was straightforward, and from the results the loads in individual spokes were deduced by integrating radial forces in the disc over the appropriate arc of the rim. This was, I believe, the first use of this device and no evidence was available as to its limitations. In consequence a number of simplified wheels were built and the forces in the spokes due to rim loads were measured. Experiments on wheels with three, four and six spokes respectively showed that the analysis was accurate and that the experimental technique was satisfactory, so that it was possible to proceed with confidence to the experimental analysis of wheels with twelve and twenty-four spokes. Calculations for these were intractable by the first analysis, but quite simple using the second, or idealised disc, treatment. A comparison of calculated and experimental values of the spoke loads enabled deductions to be made as to the scope and limitations of the new method of analysis.

Other experiments were made at later dates with the object of testing the value of this method of attack – the *leitmotiv* of the book, to quote your reviewer – when applied to a variety of problems, and I believe that those experiments were well worth while.

These illustrations and explanations may serve to show that with very few exceptions the experiments cited in the book are not of the simple type that one might show a student to illustrate, say, the triangle of forces, but were designed to explore points not readily amenable to theoretical analysis.

In conclusion, I suggest that, apart altogether from the purposes which I have outlined, there is a sound psychological objective in an experiment which demonstrates the validity of an engineering calculation even though the assumptions and treatment upon which it is based are in themselves entirely convincing to the mathematical mind. Many engineers are unwilling to apply a new method based on mathematical analysis alone, but if the results obtained can be reproduced experimentally they are much happier in its acceptance and use. Since the primary function of the engineering scientist is to help those engaged in the practice of the profession, any legitimate means of emphasising the value or, better still, the necessity of scientific approach, should be exploited. In my view, most types of experiment provide valuable means of increasing co-operation between the scientist and the practitioner by inspiring confidence. It will be unfortunate if the laboratory does not maintain its position in engineering research and development.

A. J. S. PIPPARD

London

SIR – Your review of Professor Pippard's *Studies in Elastic Structures* and his own letter on the subject, in your issue of September 18th, encourage comment on the general question of the need for experimental work in engineering. In a broad and perhaps rather flippant sense every new engineering design is an experiment, as small departures from convention may have disproportionately disastrous results. If the product (a machine or structure of some kind) is used by its manufacturer, the manufacturer does the experiment. If he sells it to a user, then the user does the experiment. There is nothing essentially unethical about this, as the only reliable experiment may be actual use, nor is it certainly wrong for the user to be allowed to do the experiment unwittingly, as use under specially observed conditions may give a misleading idea of what is likely to happen in normal service. (A locomotive superintendent once said that blue painting the chimney of any locomotive would reduce its coal consumption by at least 10 per cent, because its distinction would suggest that it was under special observation and so induce the men to work it properly.)

Acceptance of the principle of experiment by the user becomes easier when it is reflected that where very large plant is concerned no other course may be possible; experiments on a model are useless where the main experimental feature is that of size. None of the three manufacturers of boiler, turbine and alternator in a large electric power generating plant may be able to experiment on his own product because of lack of equipment for producing or dissipating the power to be handled. Experimental features in any one of the main components can be thoroughly tested only when all three are brought together, and this may be impracticable anywhere but on site. It may be argued that as separate experiment is impossible where it would be most valuable, why consider it in any case? There is some force in this where a 'one off' job is concerned, but when quantity production of a small item is contemplated, then preliminary experiment by the manufacturer is practicable and financially justifiable. These remarks refer, not to a product of completely experimental character, but to one of accepted general design but with experimental features. Anything entirely or largely novel will be recognised as such by the user, who will either agree to use it to the best advantage whilst allowing experimental data to be gathered from it, or will insist on thorough tests before he takes it over.

Tests on any plant in service are likely to be useful, but unlikely to produce as much information as if they were run in conditions devised solely to facilitate experimental work. Consequently it is technically preferable, where commercially possible, to carry out experiments on a pilot model, or even on the actual product before setting to work. If it is expected that the experiments will produce information useful in design of many subsequent items of generally similar character, then they may justify the building of plant solely for experimental purposes if the alternative of calculation is impracticable or deemed to be unconvincing. An example of this is the classic series of tests made by Professor Goss, of Purdue University, with the object of determining the best proportions of the draught-producing elements in a locomotive smokebox. It is safe to say that no one suggested that that end could be achieved by calculation and, judging by the variety of designs actually used before (and since) the publication of the Goss results, few designers approached it by guesswork. Recommendations based on examination of the observations made on the performances of many different combinations of important dimensions, were expressed in simple working formulae for designers. This therefore was a case of carefully controlled experiments on a machine reserved for the purpose with the object of rationalising the method of design of a fundamentally important feature of the steam locomotive. Another example of observations made to discover design data is the measurement of twist in a multi-throw crankshaft under known torque. The figures thus determined for torsional stiffness are used to estimate the values of certain constants in a stiffness formula of rational algebraic character. A knowledge of the torsional stiffness of the crankshaft is required in predetermining the critical speeds for torsional vibration. More recently, bending stresses produced by crankpin loads have been systematically measured by strain gauges with the similar object of establishing formulae by which the maximum stresses in any crankshaft of reasonably similar design may be calculated with a satisfactorily small tolerance.

An overdue experiment might be carried out to show (if possible) by direct test whether there is rational justification for the use of the present British Standards formulae suggesting that the permissible loading of the working surfaces of gear teeth is proportional to the 0.8th power of the radius of relative curvature of the contacting surfaces. The British Standard practice is based on the results of examining service records of a large number of gears of different materials and sizes, but it does not apparently agree with either the Herz stress analysis for dry surfaces or with the result of the Martin fluid pressure analysis for lubricated surfaces. The British Standards formulae for load capacity of gears are irrational in several different ways, but nevertheless the results they give when applied to the majority of gears of not abnormal dimensions are not dangerously or inconveniently misleading. The value of conducting a series of experiments to test any proposed rational system may therefore be questioned by anyone who uses gears for transmitting power and not merely as a topic for academic discussion. If gears are to be made in quantities for specially exacting duties, e.g. aircraft service, where expense is secondary to saving in weight, it is valuable to conduct experiments on pilot pairs. This is a case of 'testing' rather than experimenting, except that it affords opportunity for investigating the effect of changing the amount and character of profile modification of the teeth. This feature has a great influence on the maximum stresses induced in spur gear teeth, and although the effect may be estimated with fair accuracy for any assumed value of the profile modification, direct confirmation is clearly desirable where variation of a few ten-thousandths of an inch in tooth form may be vital.

Engineers are naturally reluctant to rely solely on the results of calculation in critical cases, particularly where experimental verification costs only a small fraction of the total profit to be expected if a successful design is achieved or of the possible losses if it is not. Here perhaps is the crux of the matter in commerce as distinct from science. By how much will knowledge of the result of the experiment reduce the probability of failure in the design approved in the light of that knowledge? Does the responsible engineer think that the economy by the probable reduction in failures will be greater than the cost of the experiments in money and time? The answer depends, as so frequently it does, on the psychology and experience of the individual. He may fear to suggest to those who matter to him that he lacks complete confidence in himself, and open experiment may do just that. He may decide, as Edison once did, that "he can guess nearer than the mathematicians can figure"; he may feel confident that what looks right to him on the drawing board will work satisfactorily in practice; or he may tell himself that even if it does not, he can evade serious consequences.

W. A. Tuplin

Sheffield

SIR – So far as the discussion between your reviewer and Professor Pippard has gone there appears to be a difference of opinion only on the value of experiments which reproduce the simplifying assumptions of the mathematical analysis they are designed to check, which, with an exception in favour of teaching, your reviewer considers of questionable value. He suggests, as for example, that a scale model frame test would be more informative than a test on a pin-jointed model frame if the full-sized prototype is not in fact pin jointed.

I submit that it is an over-simplification of the position to suppose that we can make a model which is absolutely representative of either the analytical assumptions or of the prototype. In all engineering problems the most careful attempt to reproduce the analytical abstractions is limited by the nature of real materials, and in particular by size effects and lack of homogeneity, by our inability to reproduce such items as the frictionless pin joint without initial clearance or permanent set due to localised loading, and by the effect of approximations in the mathematical treatment. Until it has been established that the simplified model gives results in close agreement with the analysis it is not possible to understand the discrepancies which may be found with more representative models or on the prototype.

There is a further point which requires emphasis: engineering is still to a not inconsiderable extent an art. There is a disturbing tendency abroad to assume that all engineering problems are capable of exact analytical solution, and that not only are the results of such analysis valid without physical checks, but that any problems not so soluble are not engineering. I contend that experiment is essential in all four of Professor Pippard's categories: exploratory, confirmatory, for the production of data or to replace computation in specific problems. Why should it be thought almost immoral, verging on cheating, to avoid difficult analysis by well-planned experiment to use dial gauges rather than differential operators? Personally, I would sooner trust the results of the trial and would be unwilling to pin too much faith in the analysis until checked by experiment on model or prototype.

E. MCEWEN

Duzham

SIR – Professor Pippard, in a letter to you dated September 18th, set out his views regarding the role of experiments in applied science, and expressed strong agreement with my contention that this question is important. He has given an admirable lead to a discussion which I hope will be wide and general; but here, since his letter dealt specifically with a question that I raised in a review of his book, I confine myself to that issue and maintain my anonymity. I regret that indisposition has prevented an earlier reply.

My concern, in what I wrote, was with his category (b) – Confirmatory experiments made after analysis. I specifically excluded experiments used in teaching, admitting that these can help students to comprehension of a theory. On that ground alone I would urge the need of well-equipped engineering laboratories; and equally I would accept Professor Pippard's contention that what is true of students is true of many practising engineers. All honest means to persuasion are surely justified: none-the-less, the philosophical question still remains unanswered: under what conditions, in a strict analysis, do confirmatory experiments reinforce a theoretical treatment?

As I see the matter, before that question can be answered we must have clear notions as to what disagreement would imply; and I would answer: "Not when the experiments are designed to fulfil every simplifying assumption that is made in the theory"; because lack of agreement, then, is to me unthinkable unless as resulting either from faulty experimental technique or from errors in computation. Of agreement, in the circumstances I have mentioned, two interpretations seem to be possible: either (1) the experiment has been used to verify the accuracy of computations which can be checked more easily and more reliably in other ways; or (2) what has really been established is the soundness of the experimental methods – the accuracy of the computations is assumed.

I had also meant (in the nature of the case) my remarks to be confined to structural experiments; and in this connection I add something, now, about Professor Pippard's category (c) – Experiments made in conjunction with analysis to provide essential data or to obtain empirical formulas. The strut, which he chose as an example of this class, is an excellent example of my thesis: that experiments made in parallel with theoretical treatment have real value only when they take account of factors (e.g. of rigid joints as actually fabricated) which theory is compelled to disregard, or of factors, like the variability of real materials, which are beyond the power of theory to assess. For a strut may be tested, in category (c), either (1) as a component, too complex for theoretical treatment, of a structure too large to be tested in its entirety or (2), as in Robertson's tests which Professor Pippard cites, with the aim of obtaining knowledge of a factor shown by theory to be important but beyond the power of theory to predict. In neither case, I would say, does the experiment duplicate theory; and the excellence of the example, from my standpoint, lies in the fact that (at least, within the elastic range) the theory of a strut's behaviour is almost exact – and explains the large effect of small inaccuracies which in practice are unavoidable. If the initial form of the centre-line were exactly known, there would be (for a uniform section) no need to experiment: in practice it is not known, but experiments interpreted by theory have put bounds to the imperfections which must be allowed for.

In sum, my concern thus far has been with the complementary roles of theory and experiment. I have nothing now to remark in relation to categories (a), (d) or (c), which I take to be cases in which theory is non-existent either because it presents too great difficulty or because it is not worth while.

YOUR REVIEWER

NOTE

* First published in *The Engineer* **196** (1953), 369–370, 465–466, 561.

F. V. LAZAREV AND M. K. TRIFONOVA

THE ROLE OF APPARATUS IN COGNITION AND ITS CLASSIFICATION*

Efficient investigation of the problems of the theory of knowledge from a scientific, dialectical and materialistic point of view presupposes, as V. I. Lenin stressed, dialectical elaboration 'of the history of human thought, science and technology' (V. I. Lenin, *Coll. Works* **XXIX** (5th ed.), Moscow 1963, p. 131). In this connexion it has to be mentioned that systematic analysis of the gnosiological functions and information properties of the various types of apparatus used in science, yields material of interest for the understanding of the cognitive mechanism of man, especially of the sensory stage of cognition and also of the correlation between the empirical and the theoretical levels in scientific cognition.

In our philosophical literature of recent years a certain interest can be observed in problems connected with the investigation of apparatus as a means of cognition (see the works of M. É. Omel'janovskij, B. Ja. Paxomov, Ju. B. Rumer, N. V. Topol' and others).

In this article an attempt is made to consider certain gnosiological functions and information properties of various types and classes of apparatus and on the basis of the criteria which thus emerge, to propose a corresponding classification.

I. A GENERAL CHARACTERIZATION OF APPARATUS AS A MEANS OF COGNITION

Apparatus, as a means of cognition, has been used by men since the first steps in the scientific investigation of nature. But only when modern experimental natural science came into being, the foundation of which was laid by Leonardo da Vinci, Galileo and Newton, did apparatus become an important and systematically applied device in scientific activity. As is well-known, the introduction of apparatus into the process of cognition is caused by a whole series of important factors connected with the necessity: (1) to overcome the limits of the sense organs of man; (2) to create the experimental conditions for manifestation of the physical pro-

F. Rapp (ed.), Contributions to a Philosophy of Technology, 197–209.

cess being investigated; (3) to transform the information about the object being investigated into a form accessible to sensory perception; (4) to obtain a quantitative expression of certain characteristics of the phenomenon being investigated.

What is apparatus? This term applies to a means of cognition consisting in an artificial device or a natural material product, which man in the process of cognition brings into specific interaction with the object investigated, with the aim of obtaining useful information about the latter.

The development of experimental science allows the conclusion that practically every material object can have the function of apparatus – a solid body, a liquid or gaseous medium, a crystal lattice, etc. It is evident that a particular material object has not of itself the function of apparatus, but only when it is combined with the sense organs of man and when it serves as a specific transmitter of information. What are the conditions for this combination? The interaction between the apparatus and the object investigated must lead to a state of the registering device which can be directly determined by the sense organs. This thesis is stressed by many authors (N. Bohr, M. A. Markov, V. A. Fok). It follows especially from the fact that man himself as a cognitive being "physically, as a device of investigation, is a macroscopic apparatus" (M. A. Markov, 'On the Nature of Physical Knowledge' [in Russian], *Voprosy filosofii* **1947**, No. 2, p. 152). This can be explained by the fact that historically, during the process of evolution, man had to deal in practice with macroscopic objects. Although the thesis under consideration is fundamental, it is nonetheless not to be taken absolutely. Whilst our perception is evidently *always* macroscopic (at least for contemporary man), the question of the macroscopic character of sensation needs to be treated more precisely.

The application of a particular piece of apparatus in scientific research is always connected with the solution of a certain task of cognition, with the necessity of obtaining new information. This information can be of two kinds: a qualitative kind (especially information obtained as a sensory image by means of the sense organs) and a quantitative metrical kind. The apparatus used in this connection can be classified provisionally under the two headings of qualitative and quantitative.

The apparatus of the first class is introduced into the cognitive activity in cases where the investigator is interested in information about the qualitative aspect of the object (its structure, geometrical form, colour, smell, etc.), when this information cannot be obtained directly by means of the sense organs owing to their limits (especially because of the existence of a threshold of sensations).

The basic cognitive function of the apparatus of this class consists in a maximal amplification and extension of the cognitive capacities of the sense organs. Depending on the function that a particular piece of apparatus of the first class fulfils, it can be assigned to one of three types: (1) amplifier (A-1.1), (2) analysers (A-1.2), and (3) transformers (A-1.3). Let us consider each of these types in detail.

II. AMPLIFYING APPARATUS

Apparatus A-1.1 fulfils its function of extension and amplification of the cognitive capacities of the sense organs either by *raising the exactness of observation* (for example, the microscope), or by the fact that only by means of the apparatus does the observation become *possible* at all. Generally the precision of an observation is determined by the smallest interval, within which the elements of an observed structure can be distinguished by means of the given apparatus.

The human sense organs are suited to produce an image from signals of a strictly determined kind: the eyes from light signals, the ears from acoustic signals etc. (Here we shall not treat problems like the so-called reaction to 'non-adequate stimulations'.) The signals of each kind give us certain specific information, which is generally called a sensory image (for example, a visual image). The signals belong to one and the same qualitative category, if they give rise to sensations of one and the same kind.

Apparatus A-1.1 is applied in cases where the signals coming from the object remain under normal conditions outside the threshold of sensation or where the specific character of the medium makes their direct manifestation difficult. It is evident that the action of the apparatus on the signal changes in the latter only its character as a vehicle of information. In other words, the amplifying apparatus must change the signal in such a way that it becomes accessible to the corresponding sense organs, the information supplied by the signals remaining unchanged. Hence the technical task of the apparatus A-1.1 is to convey in any way possible signals from the object investigated to the sense organs, without changing the qualitative pattern of the output signal as compared with the input signal. Therefore it is natural that the output signals of the apparatus A-1.1 produce in the observer sensations of the *very same kind* as the sensations which would arise if the signals came directly from the object to the corresponding sense organs.

Whatever type of apparatus man has to do with, in the end he receives the information in the form of a sensory image. But depending on the type of apparatus used the gnosiological 'status' of this image can vary.

What are the specific gnosiological traits of the image received by means of apparatus A-1.1? Firstly, such an image originates as a result of the direct action of the output signal on the corresponding sense organs. Furthermore, as the qualitative structure of the input signals, and, accordingly, of the output signals as well, is not changed, the nature of the sensations which form the image is preserved too. From this one can conclude that the image arising, i.e. the manifestation, taken *as a result*, is the *direct* sensory manifestation (copy) of the object investigated. But the arising of the image, i.e. the *process* of manifestation, is mediated by the apparatus used (this one can call mediation of the first order).

Thus an important characteristic of the apparatus A-1.1 consists in the fact that when applying it in the process of cognition, man obtains in every concrete case an image which, if it is taken as the final result of the manifestation, *preserves* the gnosiological status of a direct sensory image of the object investigated. From this it follows that the picture of the phenomenon which the observer produces by using apparatus A-1.1 (for example, the structure of cells under the microscope) can be described without any reference to the apparatus itself. In other words, the specific traits of the combination of apparatus of this type with the sense organs of man (and hence its involvement in the cognitive activity) are of such a kind as to produce the objective preconditions for the elimination of the apparatus from the final result of cognition. This is because the apparatus A-1.1 only overcomes the physiological limits of the sense organs, it acts as a kind of *extension* of the latter, as an 'improvement' of them. Increasing the visual power of the eye by means of the modern electronic microscope allows objects to be seen which are 200000–300000 times smaller than a pea. It was in this way that unique scientific information was obtained about the mechanism of the penetration of the living cell by certain types of virus (especially the viruses of poliomyelitis the size of 200 Å). But the possibilities of 'improvement' of the sense organs by means of apparatus are not unlimited. Apparatus improves in our sense organs only what the laws of nature allow. Therefore a certain natural limit exists in the very principles according to which the apparatus works. Among the principal limitations on the precision of apparatus are, for example, the limitations imposed by the first and second laws of thermodynamics, the quantum laws of the interaction of radiation and matter, etc.

There exists, however, a fundamentally different kind of extension of the capacities of the sense organs than that provided by the apparatus A-1.1; now we turn to consideration of this kind.

III. ANALYSING APPARATUS

The possibility of utilizing artificial or natural material products as means of cognition is connected with a number of universal traits of nature itself. According to materialistic dialectics interconnection and interaction are the basic form of being of the things and phenomena of the objective world. The emergence of certain properties of an object depends directly on the character of its relations to other objects, and on the interaction with the material medium surrounding it. "About bodies" – F. Engels wrote – "outside motion and outside all relations to other bodies, nothing can be said" (K. Marx and F. Engels, *Works* **XXXIII**, p. 67). It is evident that not every interaction between things can by itself become a means for real cognition of them. In order to transform the medium into an effective means of cognition, it is necessary to learn in every specific case to control and evaluate the character of the interaction between this medium and the object of investigation, and this is impossible without to a certain degree determining more precisely the material situation in which the investigation is performed. But in order to arrive at this more precise determination answers must be found to the following questions: (1) Which concrete property of the object is being investigated? (2) Can that aspect of the phenomenon in which the observer is interested reveal itself in the given medium with sufficient preciseness? (3) Which component of the medium is decisive for the revelation of the property being looked for and what is the influence of the other components? In so far as when one object acts upon another the latter exhibits not all, but only one or some of its properties, from this point of view the skill of an experimenter consists in his capacity to manipulate in a particular experimental situation the environment of the object in such a way that only one aspect will be decisive. In this case, according to an observation by the Soviet physicist Ju. B. Rumer, some single specific property of the object is expressed distinctly, and the given environment can be used as apparatus for the detection and investigation of this property (see Ju. B. Rumer and M. S. Ryvkin, 'On the Methodology of Physical Cognition', in *Some Patterns of Scientific Knowledge* [in Russian], Novosibirsk 1964, p. 104). It is precisely this type of apparatus that the analysing apparatus A-1.2 belongs to.

In contrast to the apparatus A-1.1, the apparatus A-1.2 does not amplify the physiological capacities of the sense organs, but extends their cognitive functions by obtaining additional information about the object, which becomes possible due to the special way in which the interaction between the apparatus and the object is organized.

The active influence on the object in the process of investigating it is a very important characteristic trait of apparatus of this type, to which belong, for example, the spectroscope, and the chromatographic paper, which decomposes a mixture of liquids into its original components. The apparatus A-1.2 exists in two varieties: (a) the analysers proper (such as the prism, which decomposes light into its spectrum); (b) indicators, which, as result of the interaction with the object investigated, allow new properties of the object to be detected (for example, a fluid medium, which allows the solubility of a particular substance to be ascertained).

Application of the apparatus A-1.2 is not due to the fact that the capacities of the sense organs are limited, but rather to the fact that the latter cannot be involved in the process of cognition without an active transformation of the object under investigation. The point is that it is necessary to apply apparatus of this type because of the specific character of the object under investigation with respect to the given cognitive task. The function of the apparatus A-1.2 does not involve any kind of change in the signals coming from the object: it transforms the object, as the source of the signals, into such a form as makes it possible to obtain by means of the sense organs new, additional information. In this case the apparatus permits the inner structure of the object to be revealed and to be represented isomorphically in a corresponding form.

It is the technical task of apparatus A-1.2 to obtain at the output the necessary information by means of direct action on the object investigated (particularly by mechanical, physical or chemical decomposition of it). From this it is evident that the application of apparatus A-1.2 presupposes as essential reference to the *level* or interval of investigation of the object. The apparatus allows an analysis only under the condition that the *qualitative pattern* of the aspects or elements forming the observed whole is preserved. In as far as ultimately this whole is the object of investigation, the information obtained about the elements or aspects and properties of the whole, will generally not make sense in itself, but only in connection with the original object. But this connection presupposes reference to the *conditions* under which the information was obtained. But evidently the apparatus itself is an important part of these conditions. As a consequence, a specific mediation of the sense data received takes place, which can be called mediation of the second order. This fact also contributes to determining the gnosiological status of the sensory image obtained by means of the apparatus A-1.2. Although the image itself arises as a result of the direct action of the output signal on the corresponding sense organ, its relation with the original object is an indirect one, which presupposes certain conditions connected with the synthesizing activity of the cognitive subject. It is also evident that the nature of sensations reflecting the investigated whole cannot coincide with that of sensations reflecting single aspects or elements of the whole. Thus, when carrying out visual observations of some chemical compound in a specific medium, we can reveal the properties of the compound investigated, the information about these properties reaching us by means of the auditory organ (hissing, crackling) and the sense of smell (odour) etc.

From this one can conclude that the picture of the phenomenon investigated which the researcher creates by means of the apparatus A-1.2, makes it necessary to take into account also to a certain degree the contribution which the apparatus introduces into the final result of the cognitive process.

IV. TRANSFORMING APPARATUS

The apparatus A-1.3 is designed for the study of a class of phenomena, the objective properties of which are such that information about them can never be gained directly by means of the sense organs (nor by means of the apparatus A-1.1 and A-1.2) without a *qualitative transformation of the vehicle of information* (for example, an electromagnetic field, infra-red radiation, ultrasonic, etc.). Apparatus of this type is also used in cases in which the phenomenon investigated is perceived by a certain sense organ with a preciseness insufficient for the aims of cognition, and in which the application of the apparatus A-1.1 or A-1.2 would be either ineffective or laborious. An example of a piece of apparatus with this function is the thermoscope, invented by Galileo at the beginning of the 17th century. In order to obtain information about such phenomena as the electromagnetic field, radiation, etc., it is necessary to find or create artificially a material object possessing a property which changes in a characteristic way under the influence of the object investigated. This *change* must possess the following properties: firstly, it must be directly perceptible to the sense organs; secondly, one must be able to judge the object of investigation according to this change.

A special instance of apparatus of this type is detecting apparatus, which has the function of providing knowledge about the presence or absence of the phenomenon being looked for in the medium under investigation.

The data from the apparatus A-1.3, on the basis of which the experimenter judges the property or phenomenon investigated, are the final link in the cause-and-effect connection between object and apparatus. It is presupposed that the connection between cause and effect is a one-one relationship, i.e. changes in the apparatus are strictly related to a precisely defined class of phenomena, which cause these changes. It is evident that the data from the apparatus (the effect) do not interest the observer in themselves as a sensory image of the registering device, but only as signals carrying information about the object investigated (the cause).

The sensory image which the observer receives, by means of detecting apparatus, is only a signal of the presence or absence of another object, i.e. it gives us information according to the principle 'yes'-'no'. As a rule, when we perceive a phenomenon directly, we receive much surplus information; but when a phenomenon is reflected by means of the apparatus A-1.3, surplus information is excluded and thus the sensory image which corresponds to the object has, as N. V. Topol' puts it, the function of a 'sensory abstraction'.

Under what conditions can some natural object be used as apparatus A-1.3? The interaction between the apparatus and the object investigated can be effectively used for the purposes of cognition, only if preliminary information about the properties and the principles of action of the apparatus is at hand, like the so-called title data (see K. Sul, *The Bubble Chamber – The Measurement and Interpretation of Data* [in Russian], Nauka, Moscow, 1970). In fixing the changes which take place in the apparatus during the process of the experiment by observing the indications of the registering device, the scientist obtains sense data material the meaning and the sense of which he can decode only by relying on the information he already has about those causal relations and laws which form the basis for the working of the apparatus. The obtaining of information by means of the apparatus A-1.3 is achieved through an inference of the cause from the effect. In other words, the information which the observer obtains as evidence from the apparatus has a conditional character. It presupposes the assumption of two premises: (1) the authenticity of the physical hypotheses on which the construction of the apparatus is based, and (2) the technical accuracy of the apparatus.

Generally speaking, the second premise is presupposed in all cases where apparatus of any type is used. But for transforming apparatus it is of particular importance, due to the special gnosiological status of the sensory image obtained by means of the apparatus A-1.3.

Inaccuracy of apparatus of the first two types can often be detected from the character of the evidence yielded by the apparatus, especially on account of the surplus information about the data obtained by the observer. In contrast to this, in the apparatus A-1.3 the signals of particular characteristics of the object investigated do not create any sensory image of the object – although they are perceptible to the senses – and therefore they do not yield any additional information, on the basis of which one could judge the correctness of the evidence from the apparatus. The sensory information in relation to the object is mediated by the premises accepted; this one might call mediation of the third order.

In the majority of cases where the apparatus A-1.3 is used, we encounter situations in which it is not possible to describe the nature of the phenomenon investigated without mentioning the apparatus. It was apparently in analysing apparatus of this type that M. A. Markov wrote: "The apparatus enters into the definition of the phenomenon. For example, the concept and the definition of the electromagnetic field necessarily involve mention of the test charge: 'the strength of the electric field is the force acting on a single test charge...' " (M. A. Markov, 'On the Nature of Physical Knowledge' [in Russian], *Voprosy filosofii* **1948**, No. 2, p. 153).

V. APPARATUS OF THE SECOND CLASS

In experimental science and in technology there exists a large class of problems the solution of which presupposes the obtaining of qualitative (metrical) information from the object. For this purpose one uses apparatus which gives at the output evidence in the language of numbers. In principle, apparatus of any of the types treated above can be supplemented by an appropriate device and thus become quantitative apparatus (apparatus of the second class). The basis for the obtaining of metrical information is the operation of counting or measuring (and also comparison, as, for example, in the use of a photometer).

When it is necessary to obtain a quantitative characterization of a *continuous* variable measuring apparatus is used. In contrast to counting, which is based on the presupposition of the objective atomism of a particular object, measurement presupposes the selection of the smallest distinguishable interval of quantification. This selection is determined by the physiological capacities of the observer (taking into account especially the capacities of the amplifying apparatus), the conditions of the task of cognition and also the character of the output device of the apparatus. The preciseness of the apparatus is, naturally, directly dependent on the graduation unit (the size of the 'quantum').

In the broader meaning of the word, one can understand by measurement any experimental procedure which permits metrical information to be obtained. Measurement, in this broad sense, presupposes the solution of two basic problems:

(1) It must be proved that it is correct to ascribe numerical values to the objects measured;

(2) It must be made clear in what sense these values are unequivocal.

In the first case, as is stressed in contemporary theories of measurement, it must be proved that any empirical system which is studied with the aim of measuring given properties of the elements of a certain field is isomorphic or homomorphic with the numerical system selected to correspond to it.

The second problem is connected with the fact that in the measuring often a great difference exists between the interpretation of the numerical results which are obtained by different kinds of measurement. This is why, to make a value unequivocal, it is always necessary to mention the type of scale according to which the measurement was carried out. It is evident that to mention the type of scale is at the same time to mention the interval of abstraction within which the given procedure of measurement has an objective meaning. In other words, this makes it possible to indicate the ontological and gnosiological presuppositions for an unequivocal interpretation of the results obtained.

The construction of a measuring instrument which guarantees correct measurements corresponding to certain original or derived scales, necessarily involves application of empirical laws and theories, including the theory of direct and indirect measurements. If the accuracy of the instrument is confirmed, then the values obtained by direct or indirect measurement are accepted as they are.

The problem of making instrumental measurements unequivocal is more complicated, because the type of scale according to which the measurement values have been obtained, cannot be established by relying only on knowledge of how the scale of the instrument is divided or even how the unit of measurement is fixed. The type of scale is determined by physical and mathematical assumptions which form the basis of the corresponding measurement procedure.

VI. APPARATUS OF THE THIRD CLASS

According to the classification suggested here, apparatus of the third class comprises the so-called measuring information systems (M.I.S.). Traditional quantitative apparatus is generally designed for measuring the value of only one variable at once. Therefore such apparatus is not suitable when it is necessary to obtain quick information from the object on the basis of the simultaneous measurement of a great number of parameters. M.I.S. are used in these cases with the aim of obtaining metrical information directly from the object of investigation, and as a rule combine the operation of measurement and control.

Both of these operations one can describe as theoretico-informational methods (see V. I. Rabinovič and M. P. Capenko, *Informational Characteristics of Means of Measurement and Control* [in Russian], Energia, Moscow, 1968). Since the manifestation of results in measurement or control is affected by chance, it is justifiable to regard them as chance events, and the (measuring or control) experiment itself as a situation in which they come about. As is well-known, in the theory of information situations are analysed in which the occurrence of a particular possible event cannot be predicted unequivocally. To give as complete a description as possible of such a situation means to indicate for each of the events the probability of its occurrence.

In order to obtain measurement (control) results a necessary condition is to perform an operation of *comparing* (whereby the functional values of the input variable and of the error are compared with the values of a previously established standard variable). In this case the amount of information makes it possible to determine how fully one can assess the values of the input variable from the results of a given experiment. To calculate the amount of information the laws of probability distribution are applied, by means of which the probability properties of the input variable and of the error are described. In the formula for the amount of information the range of the apparatus and the unit of the scale are also taken into account.

The M.I.S. generally transform signals of one physical kind into another. Depending on the cognitive task and on the needs of the user, the output information can appear in very different forms: in the shape of graphs, concrete numbers, etc. Herein lies the specific character of the cognitive status of the sensory image obtained by means of the M.I.S.

The classification of apparatus suggested in this article is only tentative, and furthermore, it only refers to apparatus, so to speak, in its 'pure form'. In the actual practice of cognition and in technology the apparatus used consists in the majority of cases of various combinations of types and classes of apparatus and complicated experimental equipment. Nevertheless, we think that in the classification given here a number of general criteria of an informational and gnosiogical order are presented, which will have to be taken into consideration when more detailed classifications are carried out. We would also mention that we certainly do not insist on the terms chosen for the classification ('amplifier', 'analyser', etc.); particularly as nearly all the terms are already applied in literature with a more narrow meaning (this does not hold for 'M.I.S.'). It was for this reason that we tried, as far as possible, to replace terms by numerical designations.

A final remark. The classification presented does not include apparatus designed for the investigation of processes connected with quantum effects. Apparatus of this type is a subject for special investigation.

Kaluga

NOTE

* Originally published under the title 'Rol' priborov v poznanii i ix klassifikacija', in *Naučnye doklady vysšej školy: Filosofskie nauki* **1970**, No. 6, 82–89.

SELECT BIBLIOGRAPHY

The following bibliography is by no means exhaustive. It concentrates rather on relevant and/or creative contributions to the problems treated in this book. Where there are various presentations of the same ideas we have attempted to select the clearest and most developed ones. The annotations are intended to afford additional information only about the relevant material. No work in the anthology itself is cited again here.The transliteration system used is that of the journal *Studies in Soviet Thought*, published by D. Reidel, Dordrecht and Boston.

I. ABBREVIATIONS

EN *The Engineer.*
PP *Periodica polytechnica – Chemical Engineering.*
PV *Proceedings of the 14th International Congress of Philosophy* (Vienna 2–9 September 1968), Vienna 1968.
PX *Prakseologia.*
TC *Technology and Culture.*
TT *Techne, Technik, Technologie: Philosophische Perspektiven* (ed. by H. Lenk and S. Moser), Pullach (Munich) 1973.

II. BIBLIOGRAPHICAL REFERENCES COVERING DIFFERENT ASPECTS OF TECHNOLOGY

Dorosiński, W., 'Wykaz ważniejszych publikacji poświęconych projektowaniu [Select Bibliography of Works on Design]', *PX* **41** (1972), 245–256. A bibliography of 266 titles, including many in Polish.
Ferguson, E. S., *Bibliography of the History of Technology*, Cambridge, Mass. – London 1968. Gives in 347 pages a comprehensive introduction to primary and secondary sources, predominantly works in English; systematic arrangement, index, annotations and frequent comments.

Gunderson, K., 'Minds and Machines', in *Contemporary Philosophy* (ed. by R. Klibansky), vol. 2, Philosophy of Science, Florence, 1968, pp. 416–425. Introductory survey of the philosophical problems of computer simulation of cognitive processes (artificial intelligence); bibliography of 42 titles.

Merrill, R. S., 'The Study of Technology', in *International Encyclopedia of the Social Sciences* (ed. by D. L. Sills), vol. 15, London–New York, 1968, pp. 576–589. Discussion of the impact of technology upon society and of conditions affecting technological change; bibliography of 127 titles.

Mitcham, C. and Mackey, R., *Bibliography of the Philosophy of Technology*, Chicago 1973 (Supplement to *TC*, vol. 14, April 1973). This is the most comprehensive survey at hand covering the various aspects of the philosophy of technology and concentrating on the period 1925–1972. Contains detailed bibliographies (non-English included). Within the epistemological studies also works on cybernetics and artificial intelligence are included. Annotation of the more relevant items.

TT, Contains a bibliography of 235 mostly German titles on philosophical problems of technology.

Cf. also the bibliographies and references mentioned in the following list.

III. WORKS DEALING WITH THE ANALYSIS OF THE STRUCTURE OF THINKING IN MODERN TECHNOLOGY

Ágoston, L., 'Zum wissenschaftlichen Begriff der Technik und über die Diskussion bezüglich seiner Definition', *PP* **8** (1964), 295–313. Gives an account of the main types of existing definitions of technology and elaborates the Marxist view of technology as a means of production.

Albring, W., 'Das Zusammenwirken von Grundlagenwissen und Technik', *Sitzungsberichte der Deutschen Akademie der Wissenschaften zu Berlin*, Klasse für Mathematik, Physik und Technik, 1967, No. 1, pp. 5–13. The interrelation between technology, physics and mathematics.

Appleton, E., 'Partnership of Science and Engineering', *EN* **198** (1954), 852–855. Adduces vacuum technique, electronics and nuclear engineering as examples of collaboration between both fields.

Beer, J. J., 'The Historical Relations of Science and Technology', *TC* **6** (1965), 547–552. Critical remarks on the theses of Price.

Bojarski, W., 'Prakseologiczny model systemu technicznego [The Praxiological Model of a Technical System]', *PX* **46** (1973), 175–197. Analysis of the process of building models of technical systems; the conformance of the system and its model to given criteria; discussion of various ways of treating the environment; description of the technical system as a multi-channelled transformer of flows.

Bunge, M., *Scientific Research* **1** and **2**, Berlin–Heidelberg–New York, 1967. An extensive treatise on the procedures of any kind of scientific research, consisting of an account of the basic strategies applied and an analysis and explication of the relevant conceptual framework. The presentation follows the general pattern of problem solving: putting forward, applying and testing scientific ideas. Many concrete examples.

Chestnut, H., *Systems Engineering Methods*, New York 1967. A broad overall systems engineering approach consisting of: introduction of the various functions and evaluations, organization of the steps to be performed, formulating and structuring the system, system evaluation for optimal decision-making, summary of tools and methods. Bibliography of 64 titles.

Clausen, H., 'Science and Design in Engineering Education and Practice – No. II', *EN* **211** (1961), 896–898. Treats design as the most important component of research and development work; cf. also Francis and Leyer.

Compton, W. D. (ed.), *The Interaction of Science and Technology*, Urbana, Ill., 1969. A collection of eight lectures dealing with the arguments raised by Price. All authors conclude that basic science does strongly interact with technology and that this interaction takes place in the modern institutions of applied science.

Conant, J. B., *Modern Science and Modern Man*, New York 1972. Gives a pragmatist interpretation of science and technology.

Dimitrieva, M. S. and Ovčarov, O. F., 'Vzaimodejstvie émpiričeskogo i teoretičeskogo v texničeskix naukax v processe progressivnogo razvitija nauki i texniki [The Interaction of Empirical and Theoretical (Elements) in the Technological Sciences in the Process of the Progressive Development of Science and Technology]', in *Problemy dejatel'nosti učenogo i naučnyx kollektivov* (ed. by S. A. Kugel' *et al.*), vol. 5, Leningrad 1973, pp. 102–105. The ways in which theoretical knowledge penetrates modern technology.

Elek, T., 'Über die Wechselbeziehungen zwischen technischen Wissenschaften, Naturwissenschaften und Mathematik', *PP* **9** (1965), 237–252. Analysis of the various aspects of the relation between engineering sciences, natural sciences and mathematics from a dialectical-materialistic point of view.

'Engineers and Scientists' (Anonymous leading article), *EN* **197** (1954), 714–715. Emphasizes the stimulation of pure science by the inventiveness of practically-minded engineers.

Feibleman, J. K., 'Pure Science, Applied Science, Technology, Engineering: An Attempt at Definitions', *TC* **2** (1961), 305–317. The differences between the various fields are based on distinct goals and interests.

Feibleman, J. K., 'Technology as Skills', *TC* **7** (1966), 318–328. Discusses the role of proficiency in the use of tools.

Finch, J. K., 'Engineering and Science: A Historical Review and Appraisal', *TC* **2** (1961), 318–332. Concludes that up to the present science has only exercised a restricted influence on technology.

Fishburn, P. C., 'On the Prospects of a Useful Unified Theory of Value for Engineering', *IEEE Transactions* **SSC-2**, No. 1 (August 1966), 27–35. Investigates the prospects for a unified theory of value useful for engineering and discusses as an alternative the possibility of a useful metatheory; 45 references.

Fores, M., 'Price, Technology and the Paper Model', *TC* **12** (1971), 621–627. A critique of the views of Price concerning the relationship between science and technology. He neglects marked differences between both fields.

Francis, A. J., 'Science and Design in Engineering Education and Practice – No. I', *EN* **211** (1961), 848–852. The innovating character and the various stages of the engineer's work; cf. also Clausen and Leyer.

Freedman, P., *The Principles of Scientific Research*, Oxford 1960. A non-technical account, based closely on practice, of the procedures relevant in planning and performing research processes, designed to inform young people just entering the field of research how to do their work.

Gasparski, W., 'Prolegomena do metodologii projektowania [Introduction to Design Methodology]', *PX* **41** (1972), 5–22. Analyzes the general characteristics of design methodology in terms of pragmatic (i.e. praxiological) and apragmatic (i.e. logical) elements.

Gibson, R. E., 'A Systems Approach to Research Management', in *Research, Development and Technological Innovation* (ed. by J. R. Bright), Homewood, Ill., 1964, pp. 34–57. Analysis of the various components of modern technology, their interactions and their feedback cycles.

Gilette, D., 'Systems Engineering: Planning', in *New Methods of Thought and Procedure* (ed. by F. Zwicky and A. G. Wilson), Berlin–Heidelberg–New York, 1967, pp. 73–84. A survey from the practitioner's point of view; cf. also Shea.

Gilfillan, S. C., 'The Social Principles of Invention', in *Research, Development and Technological Innovation* (ed. by J. R. Bright), Homewood, Ill., 1964, pp. 93–98. An account of the nature of invention and of the factors on which it depends.

Gregory, S. A. (ed.), *The Design Method*, London 1966. In 35 papers from different fields of technology a review of work and thought on the design process is given, together with an appraisal of systematic methods, and an analytical approach to the process of design, with the possibility of its extension by scientific research. Glossary of definitions; substantial bibliography of 374 titles.

Gruender, C. D., 'On Distinguishing Science and Technology', *TC* **12** (1971), 456–463. In both fields the same empirical methods are applied; but they differ in scope and the intended applicability of their theories.

Hall, A. D., *A Methodology for Systems Engineering*, Princeton, N.J., 1962. Detailed investigation of the main recurring problems of systems engineering as a process: problem definition, goal-setting, systems synthesis, systems analysis and choice among alternative systems. References at the end of each chapter.

Hickey, A. E., Jr., 'The Systems Approach: Can Engineers Use the Scientific Method?', *IRE Transactions* **EM-7**, No. 2 (June 1960), 72–80. The methods of science are described and suggested as a guide for the systems approach and for systems engineering.

Hill, L. S., 'Systems Engineering in Perspective', *IEEE Transactions* **EM-17**, No. 4 (November 1970), 124–131. Presents a background of the evolution of the systems engineering process and suggests a detailed and comprehensive resolution; 32 references.

Hübner, K., 'Von der Intentionalität der modernen Technik', *Sprache im technischen Zeitalter* **25** (1968), 27–48. Engineering cybernetics and systems theory as a general frame of reference for the structure of thinking in technology; emphasis on the systematic search for possible applications.

Hudson, L., 'Scientific Method and the Nature of Technology', *Nature* **196** (1962), 933–935. Gives an analysis of research processes in science and technology and arrives at the conclusion that scientific method, rather than a specific body of knowledge, is the real fundamental of technology.

Jones, J. C., 'A Method of Systematic Design', in *Conference on Design Methods* (ed. by J. C. Jones and D. G. Thornley), Oxford 1963, pp. 53–73. Combines logical analysis and creative thought by means of a system of notations which records every relevant item outside the memory; keeps design requirements and solutions completely separate from each other in order to arrive at the least possible compromise.

Kasprzyk, S. F., 'On the Concept of Technology and its Relation to Science and Technic', in *Proceedings of the 15th World Congress of Philosophy* (17–22 September 1973, Varna/Bulgaria), vol. 1, Sofia 1973, pp. 321–325. The characteristics of scientific, technological, and technical work in terms of their respective aims, means, and ends.

Kedrov, B. M., 'Concerning the Synthesis of Sciences', in *Man, Science, Technology: For the 15th World Congress of Philosophy* (ed. by P. N. Feodoseev *et al.*), vol. 1, pp. 67–92. The merging of scientific and technological progress is analyzed in terms of a general typology of the synthesis of modern scientific knowledge.

Konieczny, J., 'Elementi prakseologii formalnej [Elements of Formal Praxiology]', *PX* **35** (1970), 65–131. Analysis of chains of action in terms of actor, intermediary, and subject acted upon; discussion of the action types of utilization, service, interaction, and exploitation. The relationship between two chains of action is considered as well as the environment involved. An interpretation in terms of theory of conflict, teaching theory, and production theory is given. Extensive use of logicomathematical machinery. Bibliography of 70 Polish titles.

Korach, M., 'On Methodological Problems of Technology', *PP* **2** (1958), 145–171. Detailed analysis of the methodological differences between science and technology. Characteristic of technology are: the great number of variables, the scale of apparatus and the relevance of cost. Discussion of model testing, pilot plant experiments and the role of mathematics. Many Hungarian references.

Kotarbiński, T. *et al.*, 'La justification active', in *Démonstration, vérification, justification* (Entretiens de l'Institut International de Philosophie, Liège, Septembre 1967), Louvain–Paris 1968, pp. 273–301. A paper of Kotarbiński followed by the comments of eleven authors concerning an investigation of the action structures by which propositions involving predictions or rational decisions are 'justified'.

Kotarbiński, T., 'Praxiological Sentences and How They are Proved', in *Logic, Methodology and Philosophy of Science* (ed. by E. Nagel, P. Suppes and A. Tarski), Stanford 1962, pp. 211–223. Survey of the methodological problems of efficient actions.

Kranzberg, M., 'The Unity of Science-Technology', *American Scientist* **55** (1967), 48–66. The intellectual and the social needs bringing science and technology into ever closer association.

Kranzberg, M., 'The Disunity of Science-Technology', *American Scientist* **56** (1968), 21–34. The spectrum of differentiations in the relationship between science and engineering.

Lenk, H., 'Zu neueren Ansätzen der Technikphilosophie', in *TT*, pp. 198–231. Methodological differences between science and technology. Diversification within the technological sciences and the prospects for a general theory of technology. Abundant references.

Leyer, A., 'Science and Design in Engineering Education and Practice – No. III', *EN* **211** (1961), 936–938. The creative character of design; cf. also Francis and Clausen.

Lohmann, H., 'Zur Theorie und Praxis der Heuristik in der Ingenieurerziehung', *Wissenschaftliche Zeitschrift der Technischen Hochschule Dresden* **9** (1959/60), 1069–1096 and 1281–1321. Problem-solving procedures of engineering design as determined by the elements of lacuna, analogy and adaptation; the role of language and of technological and scientific knowledge. Application to engineering education. Many German references.

Lopuxov, V. L., 'K voprosu o razvitii metodologii strukturnyx klassifikacij mašin i mexanizmov [Concerning the Development of a Methodology for the Structural Classification of Machines and Mechanisms]', in *Problemy dejatel'nosti učenogo i naučnyx kollektivov* (ed. by S. A. Kugel' *et al.*), vol. 5, Leningrad 1973, pp. 111–115. Discussion of various types of classification.

Mészáros, S., 'Methoden des forscherischen Denkens in der technischen und naturwissenschaftlichen Versuchstätigkeit', *PP* **12** (1968), 419–432. Investigation of six elements relevant in research processes: analysis, synthesis, abstraction, demarcation, induction, deduction; examples from chemical engineering.

Mitcham, C. and Mackey, R. (eds.), *Philosophy and Technology*, New York–London 1972. A comprehensive selection of readings treating technology as a general philosophical problem, speculative issues included. The articles in the first chapter deal with the conceptual structure of technology. Extensive introduction; bibliography totalling 245 titles.

Moraru, I., 'Quelques particuliarités de la connaissance dans les sciences techniques: Prémisses pour une épistemologie des sciences techniques', in *Proceedings of the 15th World Congress of Philosophy* (17–22 September 1973, Varna/Bulgaria), vol. 2, Sofia 1973, pp. 343–346. Discussion of subject matter, methods, and types of knowledge in the technological sciences.

Moser, S., *Metaphysik einst und jetzt*, Berlin 1958, pp. 231–294, 'Zur Metaphysik der Technik'; reprinted in *TT*, pp. 11–81 and (in Enlish) in *Philosophy Today* **9** (1971), 129–156. On the nature of experimentation in science and technology.

Müller, J., 'Zur Bestimmung der Begriffe 'Technik' und 'technisches Gesetz' ', *Deutsche Zeitschrift für Philosophie* **15** (1967), 1431–1449. Analysis of the 'laws of technology' in terms of systems engineering and natural sciences; their demarcation from social and historical laws. Critical discussion of Teßmann.

Müller, J., 'Zum Verhältnis von Naturwissenschaft und Technik', in *Freiberger Forschungshefte* (Philosophie und Naturwissenschaften), **D** 53 (1967), 163–170. Analysis of the common traits and the differences between science and technology; they vary in their aims, but tend to apply the same procedures.

Müller, J., 'Ansatz zu einer systematischen Heuristik', *Deutsche Zeitschrift für Philosophie* **16** (1968), 698–718. The results of a research program on the application of problem-solving procedures in systematic design: background, conceptual framework, algorithm and flow of information, and input-output analysis of the conceptual procedures. Mainly German references.

Nadler, G., 'An Investigation of Design Methodology', *Management Science* **13** (June 1967), B-642–B-655. Effective design methodology must differ from research or scientific methodology, the former being based on analysis and induction the latter on system innovation and deduction. The applications of an empirical ten-step design strategy (including function determination and ideal systems development) has proved superior to conventional strategies; 29 references.

Ostrowski, J. J., 'Essai d'une typologie métaproxéologique', in *PV*, vol. 2, pp. 540–547. Survey of the literature on the analysis of (efficient) action structures.

Portevin, A., 'Deux procédés auxiliaires d'investigation scientifique: La transposition et la méthode des gradients', in *La méthode dans les sciences modernes* (ed. by F. Le Lionnais), Paris 1958, pp. 263–278. Discussion of two basic procedures of science and technology: (1) transfer of methods and results from one field to another and (2) substitution of a spatial variation for the temporal variation of a variable; examples from metallurgical engineering.

Price, D. J. de Solla, 'Is Technology Historically Independent of Science? A Study in Statistical Historiography', *TC* **6** (1965), 553–568. Investigates publication and quotation patterns in science and technology and concludes that both fields have their own separate cumulating structures. In general an interaction takes place only through the codified knowledge taught to the previous generation of students, but not at the research front. Cf. the critiques of Beer, Compton, and Fores.

Rapp, F., 'Die Technik in wissenschaftstheoretischer Sicht', in *Neue Aspekte der Wissenschaftstheorie* (ed. by H. Lenk), Braunschweig 1971, pp. 179–185. Discussion of the prospects for constructing a philosophy of technology analogous to the philosophy of science.

Reagan, M. D., 'Basic and Applied Research: A Meaningful distinction?', *Science* **155** (1967), 1383–1386. Neither the goals of the scientist nor the applicability of his work yield an operational distinction between basic and applied research (for public decision-making). Only producing new knowledge in research processes and using existing knowledge for development of particular products should be distinguished.

Rompe, R., 'Grundlagenforschung und Technik', *Sitzungsberichte der Deutschen Akademie der Wissenschaften zu Berlin*, Klasse für Mathematik, Physik und Technik, 1967, No. 1, pp. 35–44. Fields of technology depending on basic research and on ad hoc experimentation. The results of quantum theory relevant in technology.

Ropohl, G., 'Systemtechnik als umfassende Anwendung kybernetischen Denkens in der Technik', *Werkstattstechnik (wt)* **60** (1970), 452–456 and 542–545. Investigates the systems approach to technology and suggests a comprehensive discipline of engineering cybernetics; bibliography of 57 titles.

Ropohl, G., 'Prolegomena zu einem neuen Entwurf der allgemeinen Technologie', in *TT*, pp. 152–172. Proposes a general theory of technology on the basis of an input-output scheme of matter, energy, and information.

Rumpf, H., 'Gedanken zur Wissenschaftstheorie der Technikwissenschaften', in *TT*, pp. 82–107. Detailed analysis of science and technology in terms of their subject matter, their aims, the methods applied, and the structure of their propositions. The common traits of both fields transcending their differences.

Sandor, G. N., 'The Seven Stages of Engineering Design', *Mechanical Engineering* **86** (April 1964), 21–25. A systematization of creative design illustrated by component design and complex system design; 26 references.

Saunders, O. A., 'The Scientist's Contribution to Mechanical Engineering', *EN* **210** (1960), 760–761. The complex structure of the work of the engineer. Science contributes to engineering development (1) by new principles, (2) by new materials and production processes, and (3) by showing how to improve design of appliances.

Sayre, K. M., 'Instrumentation and Mechanical Agency', in *Philosophy and Cybernetics* (ed. by F. J. Crosson and K. M. Sayre), Notre Dame–London 1967, pp. 233–261. Argues that human and mechanical agents cannot be distinguished with reference merely to their respective use of instruments.

Šemenev, G. I., 'O prirode i specifike texničeskogo znanija [About the Nature and the Specific Character of Technological Knowledge]', *Naučnye doklady vysšej školy: Filosofskie nauki* **1969**, No. 6, 40–46. Concludes from a detailed analysis that there are two kinds of concepts relevant in technology: (1) concepts of science and technology, applied in the description of natural phenomena, (2) specific technological concepts referring to the way in which these phenomena are made use of in engineering.

Shea, J. F., 'Systems Engineering: Implementation', in *New Methods of*

Thought and Procedure (ed. by F. Zwicky and A. G. Wilson), Berlin–Heidelberg–New York 1967, pp. 85–96. Describes systems engineering as a mental attitude rather than as a formal discipline; cf. also Gilette.

Simon, H. A., *The Sciences of the Artificial*, Cambridge, Mass.–London 1969. Considers imperative logic and optimal decision as problems in the theory of design.

Skolimowski, H., 'Technology and Philosophy', in *Contemporary Philosophy* (ed. by R. Klibansky), vol. 2, Philosophy of Science, Florence 1968, pp. 426–437. Introductory survey including review of works on 'machines and man' and 'the logic of technological discovery'.

Skolimowski, H., 'On the Concept of Truth in Science and in Technology', in *PV*, vol. 2, pp. 553–559. Investigation of the correspondence theory of truth in science and technology; in technology the correspondence between description and object holds good if we are able to produce objects according to our description.

Sporn, P., *Foundations of Engineering*, Oxford–New York 1964. Discussion of the differences among the roles of the scientist, technician, technologist, and engineer. Engineering as the key element in translating scientific knowledge into concrete use.

Šubas, M. L., 'Nekotorye gnoseologičeskie aspekty texničeskogo projektirovanija [Some Epistemological Aspects of Technological Design]', *Voprosy filosofii* **1972**, No 9, 28–36. The function of drawings and models in the work of the engineer is described and analysed from the point of view of the theory of knowledge. The – necessarily simplified – preconceived pattern serves as an intermediate link between the idea of the designer and the finished product.

Szymański, J., 'The Definition and Application of Tools in Scientific Cognition', *Dialectics and Humanism* (The Polish Philosophical Quarterly), autumn 1973, pp. 191–202. Survey and classification of the means of cognitive activity and the various types and functions of cognition tools. Polish references.

Teßmann, K., 'Zur Bestimmung der Technik als gesellschaftlicher Erscheinung', *Deutsche Zeitschrift für Philosophie* **15** (1967), 509–527. Defines technology and its laws in terms of dialectical and historical materialism; quotes Marxist authors. Cf. Müller: 'Zur Bestimmung...'.

Tuchel, K., 'Zum Verhältnis von Kybernetik, Wissenschaft und Technik', in *PV*, vol. 2, pp. 578–585. Mutual demarcation of cybernetics, science and technology: (1) distinction between the construction of scientific theories and engineering design, (2) cybernetical methods of cognition as different from engineering cybernetics, (3) general cybernetics as part of the philosophy of technology.

Volosevič, O. M. and Meleščenko, Ju. S., 'Texničeskie nauki i ix mesto v sisteme naučnogo znanija [The Technological Sciences and Their Place Within the System of Scientific Knowledge]', in *Metodologičeskie problemy vzaimosvjazi i vzaimodejstvija nauk* (ed. by M. V. Mostepanenko *et al.*), Leningrad 1970, pp. 242–262. Discussion of the differences and the interconnections between science and technology. The classification of the technological sciences. Natural, technological and social sciences as a triad centered around mathematics. Bibliography of 37 mostly Russian titles.

Walentynowicz, B., 'On Methodology of Engineering Design', in *PV*, vol 2., pp. 586–593. Science strives for knowledge and applies analysis, engineering design aims at realization and is based on synthesis and invention; a universal methodology which includes both fields should be developed. Reference to Nadler.

Wallace, P. J., 'What is an Engineering Detail?', *EN* **218** (1964), 887–888. Discusses the role of detail design, interrelated components and standard parts within the design process.

Wendt, H., 'Die Wechselbeziehungen zwischen technischen und theoretischen Wissenschaften', *Wissenschaftliche Zeitschrift der Technischen Hochschule Karl-Marx-Stadt* **11** (1969), Sonderheft 1, 192–226. Investigates the structure of science and technology and gives a detailed analysis of the cognition and application of various types of 'technological laws'; affords concrete examples and attempts at formalization.

Zappacosta, A. D., 'Value Engineering: An Engineer's Heritage and Discipline', *IRE Transactions* **PEP-5**, No. 3 (November 1961), 8–13. Survey of the methods and the factors relevant in a quantitative product-design appraisal which ensures that every element of cost contributes proportionally to the function of the item, thus yielding a maximum product for a minimum cost.

Zieleniewski, J., 'Why Cybernetics and the Philosophy of Technical Science only?', in *PV*, vol. 2, pp. 601–608. Analysis of the areas in which general systems theory, praxiology, cybernetics and philosophy of technology overlap; suggestions for clarification of terminology.

Zieleniewski, J., 'Remarks of a Polish Praxiologist on the Subject of a Paper by Gutiérrez', *Theory and Decision* **1** (1971), 359–368. Survey of the development of theories of (efficient) human action; defence of T. Kotarbiński's praxiological approach against the charge of being aprioristic. Bibliography of 54 titles.

Zwicky, F., 'The Morphological Approach to Discovery, Invention, Research and Construction', in *New Methods of Thought and Procedure* (ed. by F. Zwicky and A. G. Wilson), Berlin–Heidelberg–New York 1967, pp. 273–297. Efficient procedures for discovery and analysis of the structure of interrelations among objects, phenomena and concepts, esp. by means of classification and combination of the relevant parameters; examples of successful application.

INDEX OF NAMES

INDEX OF SUBJECTS